CAMBRIDGE COUNTY GEOGRAPHIES

SCOTLAND

General Editor: W. Murison, M.A.

MORAY

AND

NAIRN

T0352290

Cambridge County Geographies

MORAY

AND

NAIRN

by

CHARLES MATHESON, M.A.,

M<small>C</small>LAREN HIGH SCHOOL, CALLANDER

With Maps, Diagrams and Illustrations

Cambridge:

at the University Press

1915

CAMBRIDGE UNIVERSITY PRESS
Cambridge, New York, Melbourne, Madrid, Cape Town,
Singapore, São Paulo, Delhi, Mexico City

Cambridge University Press
The Edinburgh Building, Cambridge CB2 8RU, UK

Published in the United States of America by Cambridge University Press, New York

www.cambridge.org
Information on this title: www.cambridge.org/9781107627444

First published 1915
First paperback edition 2013

A catalogue record for this publication is available from the British Library

ISBN 978-1-107-62744-4 Paperback

PREFATORY NOTE

FOR valuable aid and suggestions in connection with the preparation of this volume I have to express my indebtedness to Professor Cooper, Glasgow University; Dr Mackie, Elgin; Mr MacGregor, M.A., Forres; and Mr Thomson, Uddingston (formerly of Glenferness Public School).

C. M.

January 1915.

CONTENTS

MORAYSHIRE

CONTENTS

NAIRNSHIRE

ILLUSTRATIONS

The illustrations on pp. 27, 32, 55, 58, 65, 71, 73, 74, 77, 82, 86 and 109 are from photographs by Mr J. D. Yeadon, Elgin; those on pp. 102, 128 and 130 are from prints supplied by Mr George Strachan, Bookseller, Nairn, from photographs by Mr E. K. Hall, Mr M. G. S. Blane and Mr G. Strachan respectively; those on pp. 11, 92 and 94 are from Davidson's Series of photographs; those on pp. 19, 20 and 80 are from photographs by Mr A. Ledingham, Grantown-on-Spey; those on pp. 5, 42, 45, 52, 62, 69, 83, 107, 113, 116 and 120 are from photographs by Messrs J. Valentine & Sons; those on pp. 61, 63 and 125 are reproduced by kind permission of the Society of Antiquaries of Scotland; that on p. 9 is from a photograph by Mr James Jack, Chemist, Rothes; that on p. 14 from a photograph by Mr George Lipp; those on pp. 22 and 28 were kindly supplied by Mr Alex MacGregor, Forres, the sketches of *Elginia Mirabilis* having been made from a plaster cast belonging to Mr Taylor, Lhanbryde; the photograph on p. 47 is reproduced by kind permission of Mr J. McS. Petrie, of Glen-Logie.

MORAYSHIRE

1. County and Shire. Morayshire.

The word *shire* is of Old English origin, and meant charge, administration. The Norman Conquest introduced an alternative designation, the word *county*—through Old French from Latin *comitatus*, which in mediaeval documents stands for shire. *County* denotes the district under a count, the king's *comes*, the equivalent of the older English term *earl*. This system of local administration entered Scotland as part of the Anglo-Norman influence that strongly affected our country after 1100. Our shires differ in origin, and have arisen from a combination of causes—geographical, political and ecclesiastical. The boundaries, though often perplexing, have in the main been determined by the geographical features. Such is the case with Moray.

Formerly the Province of Moray embraced a very much larger area, including not only Nairn but also the greater part of the modern shires of Inverness and Ross. In early times the district was governed by mormaers who frequently acted independently of, or in opposition to, their nominal sovereign. But in the reign of David I the Province of Moray came to an end as a separate historical entity, and as a sign that he had incorporated it in the Scottish kingdom, David established a priory at Urquhart and a Cistercian abbey at Kinloss. Henceforward Moray included " all the plain country by the seaside, from the mouth of the river Spey to the river of

Sketch Map of Moray and Nairn in 1840

Farar or Beaulie, at the head of the Frith ; and all the valleys, glens and straths situated betwixt the Grampian Mountains south of Badenoch and the Frith of Moray, and which discharges rivers into that Frith." From time to time this area was still further reduced.

Till the latter half of the nineteenth century Morayshire consisted of two widely detached parts, separated by a portion of Inverness-shire. In order to get rid of the anomaly, "The Inverness and Elgin Boundaries Act" (1870) transferred part of the united parish of Inverallan and Cromdale from Inverness to Moray, and parts of the parishes of Abernethy and Duthil from Moray to Inverness. The boundaries of Moray were again rearranged by the Boundary Commissioners in 1891. Bellie and Rothes, parishes formerly partly in Morayshire, partly in Banffshire, were placed wholly in the former county, and Boharm, Inveravon and Keith in the latter. Of the parishes partly in Morayshire and partly in Nairnshire, Dyke and Moy was placed wholly in Morayshire; while the detached parts of the Nairnshire parish of Ardclach were transferred to the parish of Edinkillie. The parish of Cromdale had been partly in Inverness-shire and partly in Morayshire. The commissioners left the boundaries untouched by transferring the Inverness-shire part to the Inverness-shire parish of Duthil, and restricting the name of Cromdale to the Morayshire portion.

The name Moray is supposed to be an old locative plural of the Gaelic word *muir*, the sea. If this derivation be correct, Moray means "among the seaboard men." The designation is an apt one, for the Moray Firth is an all-important factor in the history of the district. As regards the derivation of Elgin, the other name for the county, authorities are completely at variance.

2. General Characteristics. Position and Relations.

Scotland is usually said to consist of three well-defined natural divisions—the Northern Highlands, the Central Lowlands, and the Southern Uplands. But the area of which Morayshire forms a part—the lowlands on the Moray Firth—does not fall very well within any of these divisions. For meteorological, topographical, and geological reasons we may look upon this region as a separate geographical unit, which, for want of a better designation, we may call the Northern Lowlands.

Most of Morayshire enjoys a very favourable climate. Besides this the county possesses the advantage of great stretches of rich soil, the result to a large extent of the weathering of the Old Red Sandstone.

Writing in 1618 Taylor, the Water-Poet, says: "the countie of Murray is the most pleasant and plentiful country in all Scotland; being plaine land, that a coach may be driven more than foure and thirtie miles one way in it, alongst by the sea-coast."

Morayshire is bounded on the north by the Moray Firth, on the east and south-east by Banffshire, on the south and south-west by Inverness-shire, and on the west by Nairnshire. The county falls naturally into two well-marked divisions. To the north an extensive plain extends westward from the Spey, between the shore and a range of hills, for nearly forty miles, and varies in breadth from five to about twelve miles. This low-lying area is

Elgin Cathedral

diversified by a series of lower ridges, in the main running nearly parallel to the coast-line. Here lies the fertile "Laigh of Moray," a district which, for the skill shown by its farmers and the quality and quantity of its crops, compares favourably with any other part of Scotland. For centuries its productiveness has been almost proverbial. In the famine which was general all over the country towards the close of the sixteenth century the Laigh was able not only to produce sufficient for local requirements, but also to spare large quantities of corn for less-favoured parts of the land. Men came across the Grampians even from Forfarshire to buy the necessaries of life.

To the south of the plain lies a mountainous district where an unkindly soil and an unfavourable climate force the inhabitants to give their attention chiefly to pastoral pursuits.

The rivers of the county flow in a north-easterly direction. In part of their courses they flow with great rapidity, and occasionally do great damage by flooding, especially in the low-lying parts of the county.

Nearly all the important towns and villages of the county are situated in the northern plain. Elgin owes its position as capital of the county to its castle and its cathedral, and is a good example of a town nurtured into prosperity under the aegis of the church. Purely geographical considerations would have led to the selection of Burghead or, less likely, Forres as the centre of government.

3. Size. Shape. Boundaries.

The total area of the county, excluding water, is 304,931 acres. In point of size it ranks eighteenth among the counties of Scotland. Its greatest length, north-east to south-west, from Lossiemouth to the neighbourhood of Dulnain Bridge, is 34 miles ; its greatest breadth, east to west, from Croft of Ryeriggs to Macbeth's Hill, west of Forres, is rather less than 30 miles. In shape Morayshire is triangular, the base running from the south to the north-east, with the apex to the north-west.

The boundaries, though occasionally artificial and arbitrary, clearly show as a rule the influence of geographical control. Thus, for over twenty miles the Spey forms the boundary on the east, while on the west the north-eastern prolongation of the Monagh Lea Hills divides the county from Nairnshire. Starting from the north-east corner, the boundary line lies along the channel of the Tynet Burn. About a mile south of Ryeriggs it runs westward for a short distance, and then south, where it crosses the main road from Keith to Fochabers. Its course is then westward by Thief's Hill (819 feet) till it reaches the Spey about a mile south of Fochabers. It follows the river to a point about two miles below Rothes, where, running south-east for a short distance, it turns westward. After taking in a part of the slope of Ben Aigan (1544 feet), it again reaches the river, which is the march to the neighbourhood of Delnapot. Here it turns first south and then south-west along the watershed of the Cromdale Hills. The Spey from about two miles east

of Grantown is again the boundary to the south of
Tom-na-Croich, the most southerly point in the county.
Then its course may be regarded as north-westerly until
it touches the slopes of Carn Allt Laoigh, where the
counties of Moray, Inverness, and Nairn meet. From
here it holds northward, and strikes the Moray Firth
about five miles west of the mouth of the Findhorn.
Continuing northward the line cuts The Bar into two
almost equal areas, thus dividing it between Moray and
Nairn.

4. Surface and General Features.

The main physical features of the Highlands and
Islands of Scotland run approximately in lines from south-
west to north-east. The county of Moray forms an
excellent illustration of this fact. This becomes plain if
we compare the direction of the valleys of the Findhorn
and the Spey with that of the Outer Hebrides, the Minch,
Glenmore, Loch Fyne, Loch Tay, and Loch Ericht.

The surface of Morayshire shows great variety,
ranging from sea-level to over 2300 feet. The county
can be divided into two distinct parts. The north
consists of a lowland, the south of a highland region,
the two areas being separated by a range of hills whose
highest peak is under 1800 feet in height. Though the
south of the county is hilly, none of the eminences can
be fitly called mountains. Larig Hill (1783 feet), Carn
Ruigh (1784) and Carn Kitty (1711) are the chief heights.
This upland region forms an important watershed and is

drained by the Spey, the Lossie, and the Findhorn.   On the Banffshire border we come to the Cromdale Hills, with several peaks over 2000 feet high.   The highland country to the south—the " Brae of Moray "—abounds

The Doonie Lynn, Rothes

in glens and straths running from south to north and drained by the Spey and the Lossie.   Here lie the great valley of the Spey and the glens and straths of its tributary streams, and here is found some of the most magnificent highland scenery in Scotland. Several of the straths are very fertile; but an excessive rainfall and a lack of sunshine are serious drawbacks to agriculturists.

On the moors grouse and hares are abundant; and in the lower parts partridges, pheasants, snipe, and rabbits furnish excellent sport for the numerous shooting-tenants.

The seaboard plain contains "the garden of Scotland," and notwithstanding the northern exposure, forms one of the best agricultural parts of Scotland. Many circumstances, chiefly a mild climate and a rich alluvial soil, combine to give the district this enviable reputation. It is fertile and well-wooded and rises very gradually from the shore, until, at a distance of two miles inland, it reaches an average height of about 75 feet. For the next six miles the rise is less gradual to a height of about 600 feet. The whole district is highly cultivated or covered with beautiful woods.

The " Laigh of Moray," as this seaboard plain is called, is famous for the excellence of its climate. Occasionally, however, drought in the summer months causes great damage to the growing crops. In the highland part of the county, on the other hand, excessive rainfall is the agriculturist's most unrelenting foe. The contrast, in this respect, between north and south is well expressed in the old couplet :

> " A misty May and a drappy June
> Sets Moray up and Spey doun."

Slightly undulating in the east, the Laigh becomes practically dead level in the parishes of Alves and

Kinloss, where the productiveness of the area reaches its maximum.

To the west of Findhorn Bay lies perhaps the most extraordinary physical phenomenon in Scotland—the sand-hills of Culbin. From Kincorth to the sea-shore a large area is completely covered with sand-hills; and in the

The Culbin Sands

middle rise enormous mounds of sand, some over 100 feet in height. This central part once formed the Barony of Culbin. About 250 years ago the Barony was part of the most fertile and prosperous district of the county, frequently referred to as the " Granary of Moray." It is impossible to estimate the extent of arable land on the estate, but we know that it comprised a home farm, several small farms, and numerous crofts. In 1654 the valued rental of the Barony in the parish of Dyke was no less than £913. 18s. 4d. Scots. The sand encroached on the arable land for the first time in 1676 and every succeeding year made matters worse. In 1695 an Act of Parliament was passed to prevent the pulling of bents (*Ammophila arundinacea*) and so retard the progress of the sand. The Act states that : "The Barony of Culbin, and house and yards thereof, is quite ruined and overspread with sand." While local tradition is wrong in ascribing the whole catastrophe to one great storm, still on parts of the estate the advance of the sand was sudden and unexpected. Portions of the old land are exposed from time to time. In one place so exposed a ploughman was suddenly interrupted while at work, for half of the ridge is ploughed and the other half is untouched. At another place a plough was discovered after having lain beneath the sand for well-nigh two centuries.

Some believe that, owing to the continual drifting of the sand eastward, the district will one day be again free from sand. This, however, is most improbable. The breaking up of the old coast-line seems to have furnished the sand. Carried westward and cast up on the shore between Findhorn and Nairn, it was then blown inland by the strong westerly winds. Driven further and further east, it in course of time formed huge mounds, spreading itself over a tract of country eight miles in length and in some parts two miles in breadth. But the movement eastward still goes on with little or no cessation, and no one who has seen the Culbin Sands during a wind-storm is ever likely to forget the sight. After being driven into the Bay of Findhorn, the sand is carried out to sea where the inshore tidal current catches it up and deposits it afterwards on different parts of the coast. Then the westerly winds seize upon it and the cycle begins once more.

5. Watershed. Rivers. Lakes. Coast-line.

Except on the Cromdale Hills the boundary line of the county is almost entirely independent of the watershed. A straight line drawn from the mouth of the Spey to Carn Allt Laoigh is roughly the division between the two main drainage areas. The land to the south and east of this line drains into the Spey, that to the north into the Lossie and the Findhorn. From Carn Kitty streams radiate to all three rivers. It will thus be seen that the rainfall of the county finds its way to the Moray Firth and that practically the whole county is drained by three rivers and their tributaries.

The Spey, the chief river of Morayshire, rises in Loch Spey, in Inverness-shire, at a height of 1142 feet above sea-level. The Spey flows first east, then north-east through Inverness-shire, and afterwards north-east through Morayshire or on the boundary between Moray and Banff. On approaching Fochabers, it becomes wholly a Morayshire stream and falls by several mouths into the Moray Firth at Kingston.

The Spey at Fochabers

The Spey, in the lower part of its course, is the swiftest river in Scotland. Its impetuosity is due partly to the great volume of water carried into it by its numerous tributaries from the elevated region where it originates, and partly to the great height of its source above sea-level. "In the first section," says Mr Chisholm, "from Loch Spey to Laggan Bridge...the average fall of the bed is at the rate of 23 feet per mile. In the second, extending as far as Grantown...the average fall is 5·3 feet per mile, the three nearly horizontal portions in this section representing three extinct lake-basins. In the third section...the average fall is 14 feet per mile, and is very evenly distributed, so that almost throughout there is a slope of about 1 in 378." As the result of a geological examination Mr Hinxman thinks that the bed of the river is "composed of two sections of different age and at unequal stages of erosion. The lower section has been rejuvenated by uplift, and, contrary to the normal conditions of an orthodox stream, is at once younger in age and steeper in gradient than the higher portion."

The total length of the Spey is estimated at 110 miles, of which about 37 are in Inverness-shire. The only Scottish rivers that surpass it in length are the Tay and the Tweed. The area of its basin is 1190 square miles, ranking next to the Tay, the Tweed, and the Clyde. It is usual to divide the valley of the Spey into three parts—Badenoch, Strathspey, and Speyside. Badenoch lies wholly outside the county. Strathspey comes as far as the pass of Craigellachie, which is closely hemmed in by the Craigellachie Rock on the north and a spur of

Ben Rinnes on the south. Speyside is the rest of the valley from Craigellachie to the Firth.

The Spey first touches Morayshire about three miles south-west of Grantown, its volume already increased by the waters of many mountain torrents and burns. From Speybridge to Craigellachie the river follows a serpentine course towards the north-east through a highly picturesque and beautifully wooded country—a country frequented by tourists and sportsmen, and a favourite resort of the late King Edward. For beauty of scenery Strathspey has few rivals in Scotland. Before reaching Craigellachie, the Spey is joined on the left bank by the Burns of Tulchan and Ballintomb, and on the right by the crystal-clear Avon.

> "The water of Avon runs so clear,
> It would cheat a man of a hundred year."

From Craigellachie, where on the right the Fiddich enters, the river rushes on in majestic curves past Dandaleith and through the Haughs of Rothes. At Dandaleith is one of the four specially famed farms in the valley of the Spey. Every son of Moray is familiar with the lines:

"Dipple, Dundurcas, Dandaleith and Dalvey
Are the four bonniest haughs on the lang run o' Spey."

The scenery between Craigellachie and Rothes is superb. On the right are the woods of Arndilly, decking the lower slopes of Ben Aigan. Opposite are the woods of Conerock, while in the plain between the river swings onward to the sea. Flowing through the narrow pass of Sourden, where occurs probably the deepest pool in its course, the Spey increases its velocity, and pursuing a course almost due north forms a delta of ever-varying islands, and then at Kingston dashes into Spey Bay. In 1860 the river cut a new channel for its mouth—a result of the large quantity of loose stones and gravel it carried.

In salmon fishing the Spey ranks next to the Tay and the Tweed. It also yields trout and "finnock" or whitling.

Unlike the Spey and the Findhorn, the Lossie is entirely a Morayshire stream. It rises on the flanks of Carn Kitty in the small lakes Lossie and Trevie, and it also receives a stream from Loch Noir. It follows a very winding course, generally north-by-east, until it nears the Quarry Wood. It then turns east forming a right angle, which contains the town of Elgin. A short distance further on it again flows north and enters the Firth at Lossiemouth. Few woods are found in its valley, and for fishing it is inferior to the other rivers of the county.

The Findhorn, like the Spey, has its source in the county of Inverness, in the Monagh Lea Hills. Of its total length of nearly 70 miles, only the last 12 or thereby are in Morayshire. Before entering the Firth at Findhorn Bay, the river separates the Forest of Darnaway, an old royal forest, from the Woods of Altyre. For the first six or seven miles of its course in Morayshire, the Findhorn is not merely unsurpassed but actually unrivalled for beauty of scenery. Foaming torrents, towering crags, quiet mountain-sides, rich woods, gently rounded hills and highly cultivated valleys—every variety of picturesqueness, in short, is to be found in the course of this romantic river. It is a paradise for the artist and the lover of nature. Near the place where the combined waters of the Divie and the Dorbock join the Findhorn is the famous

Randolph's Leap, on the Findhorn

Randolph's Leap, where rock and wood and river combine to form a scene of wildest grandeur.

In geological interest also the Findhorn takes a foremost place. Various barriers occur in its course. The best known is at Sluie, between the Forest of Darnaway and the Altyre Woods, where at an abrupt fault may be seen the junction of the sandstone of the district and the primitive geological formations.

A great change has taken place in the course of the river. Formerly the Findhorn flowed round to the north of the lands of Binsness and then, forming a right angle, turned westward. It flowed west for about three miles

The Bay of Findhorn

and discharged into the Firth at The Bar. The alteration of its course was probably caused by the silting up of the river owing to the accumulation of sand driven eastward from Culbin. To-day it forms a tidal lagoon, about two miles long and two and a half miles broad, called the Bay of Findhorn, with a mouth less than half a mile wide.

Though still a good salmon and trout river, the Findhorn now falls far short of its old reputation. Less than a century ago 360 salmon were taken from one pool in a single day.

The Spey, the Lossie, and the Findhorn are all liable to sudden flooding. This is due to the high rainfall in the elevated parts of their drainage areas and to the low-lying nature of the Laigh lands. To this flooding in past ages may be attributed to a large extent the fertility of the soil.

Lochindorb

The lakes of Morayshire are few in number and insignificant in size. Lochindorb in the south on the margin of the lowlands between the Spey and the Findhorn is the most considerable, being about two miles long and three-quarters of a mile at its widest. On a small island, partly artificial, in the northern half of the lake, are the ruins of Lochindorb Castle, once the stronghold of the

"Wolf of Badenoch." In 1303 the castle was reduced by Edward I and in 1336 by Edward III. Among the ruins is found a peculiar plant, resembling red cabbage, known locally as "Lochindorb Kail." The country people transplant it to their gardens and use it as a vegetable.

Loch Spynie, two and a half miles north of Elgin, is estimated to have covered in 1779 an area of 2500 acres. Early in the nineteenth century it was resolved to reclaim the land by draining the loch. According to the *Statistical Account* the drainage operations cost £10,744. In the drains and canals marine shells, oysters, cockles, etc. are found, and their presence clearly indicates a connection at no very remote date between the loch and the Firth. Spynie is now a marsh of little more than 100 acres. It is the haunt of countless waterfowl. The railway from Elgin to Lossiemouth passes over the former bed of the lake.

Loch-na-bo, near Lhanbryde, is surrounded by woods on all sides and is a perfect beauty spot. Loch of Blairs, two and a half miles south of Forres, Loch Romach, three miles south of Rafford, Loch Dallas, four miles south-west of Dallas, are also worthy of mention.

The coastline of the county is remarkably free from indentation. From the eastern boundary, where the Tynet Burn enters the Firth, to the mouth of the Spey the coast is low-lying and only rarely reaches a height of twenty feet. The prevailing rocks are a dark red sandstone, which has been quarried for building purposes, and conglomerate. From Garmouth a fine view is obtained.

Eastward may be seen Buckie and the Bin of Cullen ; to the north the silted-up port of Kingston and the broad waters of the Firth ; to the west, Lossiemouth ; while inland Ben Aigan and Ben Rinnes stand as sentinels of the strath.

From Kingston to Lossiemouth the coast is low and sandy, and from the sand rises a series of bent-covered hills and pebbly beaches. Just above the mouth of the

Covesea Lighthouse

Lossie is the end of the canal which drains Loch Spynie. Formerly, and especially in the floods of 1829, low tracts of land along the Lossie suffered great damage from inundation, but embankments now serve to keep back the river when in flood. The peninsula on which stands Branderburgh, a part of Lossiemouth, contains the most northerly point in the county. Lossiemouth golf course

is one of many excellent golf courses on the southern
shore of the Firth.   About two miles west of Lossiemouth
is Covesea Lighthouse, which, owing to the level nature
of the country, is a landmark for miles around.   Covesea
Caves lie about two miles farther west.   As late probably
as the eleventh century the low-lying strip of ground
between Lossiemouth and Burghead Bay was under sea-
level.   The strong westward tidal current carrying shingle
from the coast of Banffshire, has caused a considerable
addition to the land area.   Where the coastline projected,
however, the sea has gained on the land.   Thus at
Burghead, according to tradition, a wide stretch of land
intervened between the old fort and the sea.   To-day
the remains of the fort overhang the waves.   The village
of Findhorn has suffered devastation twice.   The first
village was destroyed by the drifting of sand ; the second
was ruined by an inundation in 1701 ; while part of the
present village is threatened both by sand-drift and by the
terrible surf that beats against it.   West of Findhorn Bay
are the Culbin Sands.   About half-way between the
mouth of the Findhorn and of the river Nairn is a strip
of land called The Bar, consisting mainly of bent-clad
sand-hills.   The Bar is three miles long, and its greatest
width is about four hundred yards.

6.   Geology and Soil.

Geologists divide the rocks which form the earth's
crust into three classes : (1) *Sedimentary* ; (2) *Igneous*, and
(3) *Metamorphic*.

Sedimentary rocks form the most important group. Initially laid down in water, they have since by some movement of the earth's crust, been raised above sea-level. Such rocks are deposited in *strata* or beds, varying in thickness from an inch or less to several thousand feet. Though originally laid down horizontally, they are more frequently found tilted to a greater or less extent by subsequent movements of the earth's crust. Many sedimentary rocks contain fossils, i.e. remains of plants and animals which lived at the time when the rocks were formed. The fossils are of great value in deciding the age of the rocks and their mode of formation. This is of so much importance that those who study fossils— palaeontologists—have divided sedimentary rocks into four classes according to the character of their fossils: (1) Primary or Palaeozoic; (2) Secondary or Mesozoic; (3) Tertiary or Cainozoic; and (4) Quaternary or Post-Tertiary. Among the most important sedimentary rocks are sandstones, clays, shales, gravels, limestone, chalk, and coal. The three last are of organic origin. Coal consists of the remains of plants which flourished many thousands of years ago. The coal lies on clay which was the old soil in which the plants originally grew. Chalk and limestone are formed from the remains of animals which lived in water. When these animals died their skeletons sank and in the course of ages hardened into rocks.

Of the igneous rocks some have solidified on the surface of the earth, e.g. lava; while others, like granite, have solidified deep down in the recesses of the earth, becoming visible only when the overlying rocks were carried away by denudation. The absence of bedding is a characteristic of igneous rocks.

When the original nature of a rock is completely altered by pressure, or heat, or chemical change, the rock is said to be metamorphic. By such means sandstone may be altered into quartzite, limestone into marble, granite into gneiss, clay into slate, and coal into anthracite.

Sedimentary rocks have been divided, according to their age, into various classes and sub-classes. The oldest class or Palaeozoic includes, in order of age :

Cambrian,
Silurian,
Old Red Sandstone or Devonian,
Carboniferous,
Permian.

The Mesozoic or Secondary group of rocks includes:

Triassic,
Jurassic,
Cretaceous.

Geologically, Morayshire falls into two divisions. The hilly district in the south of the county is composed of metamorphic rocks, chiefly crystalline schists such as characterize the Central Highlands, while the plain in the north of the county consists of Old Red Sandstone and strata of Triassic age, overlaid with glacial deposits. In the Cromdale Hills, metamorphic rocks such as mica-schists, quartz-schists, and quartzite, predominate. Most of the metamorphic area west of the Spey is occupied by rocks which come under the Moine schists of the

Geological Survey. Throughout the area veins and
bosses of granite are found injected into these rocks.
The Findhorn, the Divie, and many of the tributaries of
the Spey show remarkably fine sections of these schists.
Near Grantown crystalline limestone with various admix-
tures is found. As a rule the strike of the crystalline schists
of the county is north-east and south-west, while the dip
is to the south-east.

A mass of granite, belonging to the same intrusions
as the Cairngorm granite, is found between Grantown
and Lochindorb. Resting unconformably on some of
the crystalline schists, are found representatives of the
upper and middle divisions of the Old Red Sandstone.
The middle series is represented by conglomerates, sandstones, shales, and clays, which are well shown to the
north of Boat of Bridge, in the banks of the Spey, and in
the Tynet Burn east of Fochabers, the latter locality
being famous for its fossil-fish bed. In the Tynet Burn
and also in the Gollachie Burn sections the bed containing
fish fossils is overlaid by pebbly sandstones and conglomerates, and above these at Gollachie is found a lava flow
of considerable thickness. The coarse conglomerate of
the valley of Rothes and of Findlay Seat belongs to this
division as well as some rocks occurring at other places
to the west. South-west of Elgin the Upper Old Red
Sandstone rocks overlap the middle series and lie directly
on the Metamorphic, as may be seen from sections in the
valley of the Lossie, and on the Findhorn, from Sluie
downwards.

Red, grey, and yellow sandstones with bands of

conglomerate form the strata of the upper division. They are well seen in the Findhorn between Sluie and Cothall, in the Lossie, and at Scaat Craig to the south of Elgin, while to the north of the town they can be traced from Bishopmill to Alves. Strata of Triassic age occur at Quarrywood, and at Spynie, Findrassie, and Lossiemouth, while the coast ridge from Covesea to Burghead is wholly composed of sandstone of this age. These strata consist

Gow's Castle, Covesea

of yellow and grey sandstones, and a curious band of chert called the cherty rock of Stotfield. The sandstones contain a wonderful series of reptilian fossils. One of these, a horned reptile, has never been found in any other part of the world, and has been called *Elginia Mirabilis*. This area was long supposed to consist solely of Old Red Sandstone, but the fossiliferous remains put it beyond doubt that some of the sandstones are Triassic, if the sandstone containing *Elginia* be not Permian.   Owing to their excellence for building purposes, both the Upper Old Red and the Triassic Sandstones have been largely quarried.

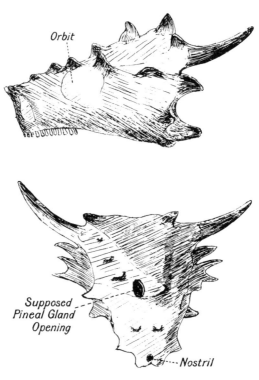

Elginia Mirabilis ($\frac{1}{4}$ actual size)

Morayshire contains extensive and important glacial deposits resulting from the action of the last Ice Age.

The deposits occur alike in the Laigh and in the river-valleys of the Brae of Moray. In the Laigh the alluvium thus formed is found mixed with the débris derived from the grinding down of Old Red Sandstone rocks. The ice crossed the low grounds in an easterly to south-easterly direction, and deposited "erratic blocks," some of which have been carried all the way from the centre of Ross-shire.

Sandy or gravelly soil is found throughout the county but especially in the north-east in the parishes of Spey-mouth, Urquhart, Lhanbryde, Drainie, and Elgin. In part of Drainie and about the town of Elgin a fertile loam prevails. Large tracts of this soil are found also in Duffus, Alves, Spynie, Kinloss, Forres, Dyke, Rafford, and Edinkillie. In the highland part of the county, the plains along the rivers and streams are covered with a sandy soil or a loam with a considerable admixture of sand. In certain parts of the Laigh there is a sub-soil of a fine sandy nature, hardened by the deposition of iron and manganese oxides, which the farmers find difficult to pierce. It is called the "Moray Pan." Luckily, however, the area where it occurs is not very wide.

7. Natural History.

Long ages ago Britain was joined to the European mainland, and the North Sea did not exist. So shallow is that sea to-day that an elevation of 100 fathoms would almost entirely obliterate it. Owing to this land connection prehistoric Britain was inhabited by the same species of plants and animals as inhabited what is now the opposite coast of Europe. After the formation of the North Sea, however, many continental plants and animals which had not previously reached Britain, found it impossible to do so. Accordingly while our country has barely forty species of land mammals, fully ninety species are found in corresponding latitudes on the continent. Every type of mammal, reptile, and amphibian that we have is found in Europe; and our only bird not seen on the continent is the red grouse of the Scottish moors.

The fauna of Moray differs little from that of Scotland as a whole. The wild cat and the marten are extinct. Badgers used to frequent the Darnaway Woods but none has been seen for over twenty years. The fox is met with; and the otter appears occasionally. Hedgehogs and moles, the common shrew, the lesser shrew, and the water-shrew are found. Four species of bats are known —the common bat, the long-eared bat, the noctule bat, and Daubenton's bat.

Rodents are numerous. The squirrel is plentiful in the wooded districts, and proves destructive to the pine-shoots in the Darnaway and the Altyre Woods. The black rat has been completely supplanted by the brown rat, which has become a pest almost everywhere. The water vole, the field vole, the bank vole, the house mouse and the field mouse are found; but the harvest mouse has never been recorded for the county. The common hare and the mountain hare occur, the former less frequently than formerly. Rabbits are plentiful everywhere, but especially where the soil is sandy. They are abundant near Forres and on the edges of the Culbin sand-hills.

The roe-deer is not so common as formerly. The red-deer occurs in the south of the county, and the fallow-deer has been introduced.

Fully 200 species of birds are recorded, and of these 98 species are known to breed in the county. The kestrel, the sparrow-hawk, and probably the merlin, are resident. The golden eagle, the common buzzard, the rough-legged buzzard, the hen-harrier, and the peregrine falcon are now rarely visible. Six kinds of owls are recorded, two of which —the tawny owl and the long-eared owl—are resident. The kingfisher and the redstart are seen occasionally on the Findhorn and elsewhere. The fieldfare and the redwing are winter visitants. The golden-crested wren is numerous in the fir-woods in winter after the autumn migrations from Scandinavia. The famous heronry of Darnaway disappeared in 1863. Swans occasionally visit Loch Spynie in severe weather during spring or autumn migrations. Wild geese of several species alight at the same loch on migration. In winter eider-ducks visit Findhorn Bay. Stock-doves were first observed in the county in 1883 and have since increased greatly in number, many pairs breeding in the rabbit-holes of the Culbin sand-hills. Among the game birds of the county are the red grouse, partridges, pheasants and capercailzies. Quails have been found in the neighbourhood of Forres and Duffus. Common, arctic, lesser, Sandwich, and roseate terns occur as summer visitants. Black-headed gulls are abundant, their favourite breeding haunts being Loch

Spynie, the lochs in the Altyre and the Darnaway Woods, and a tiny loch at the eastern end of the Culbin Sands. Among the commonest birds in the county are house-sparrows, thrushes, blackbirds, skylarks, chaffinches, robins, hedge-sparrows, lapwings, common gulls, green-

Spynie Palace and Loch

finches, yellow-hammers, blue-tits, common linnets, starlings, rooks, and wood-pigeons.

The insect life of the county shows great variety. Butterflies are numerous. The Red Admiral is found, and two specimens of the exceedingly rare Camberwell Beauty have been captured on Cluny Hill, Forres. Of the larger moths the Death's Head and one or two species of hawk-moths occur.

Nearly all the British species of reptiles and amphibians are found. The adder, the only poisonous reptile in our country, haunts many parts of the shire, including the Darnaway Woods. The viviparous lizard has been met with on the Culbin Sands, and the slow-worm or blind-worm frequents the banks of the Findhorn. Frogs and toads abound. Palmated newts are common, while the great warty and the smooth newt have been recorded from near Forres.

The flora of the county is very varied. The seashore plants include sea pink, stonecrop, scurvy-grass, kidney vetch, sea purslane, sandwort, sea plantain, and wild thyme. On the waysides may be seen creeping buttercups, dandelions, Germander speedwells, harebells, ribwort plantains, hedge mustards, hedge parsleys, tufted vetches, meadow vetchlings, knapweeds, common sorrels, St John's worts, bird's-foot trefoils, silverweeds, and yellow rattles. In May and June parts of the county are ablaze with whin and broom. Specially noticeable patches are found west of Mosstowie Station, south of Chapelton and Sanquhar, and on a ridge between Duffus and Covesea. The bogs and peatmosses are the haunt of the broad-leaved orchis, the spotted orchis, the fragrant orchis, the lousewort, the bog asphodel, the spearwort, the ragged robin, the bitter cress, the water bedstraw, and the marsh thistle. The river banks are covered with hogweed, valerian, water avens, meadowsweet, willow-herb, tussilago, wild angelica, and lady's-

mantle. In the woods the following plants abound:
anemone, wild hyacinth, stone bedstraw, common speedwell, wood-sorrel, foxglove, common tormentil, rose
campion, dog violet, cow-wheat, whortleberry, herb-
Robert, common bugle, wood-sage, bramble; and, in
the fir woods, chickweed wintergreen. The corn fields
contain yellow marigold, as says the popular rhyme,

> "The gule, the Gordon and the hoodie-craw
>   Are the three warst things that Moray ever saw";

mayweed, field mustard, field spurrey, common poppy,
blue cornflower, hemp-nettle, chickweed, field mint,
fumitory, bugloss, field lady's-mantle, and climbing bistort,
while the pastures show a profusion of buttercups, daisies,
scorpion-grass, ragworts, spear thistles, selfheals, white
clovers, and sheep-sorrels. In the waste places thrive
shepherd's-purse, groundsel, dandelion, purple dead nettle,
plantain, dock, stinging nettle, nipplewort, thalecress,
common knotgrass, chickweed, goosefoot, and pearlwort.
The commonest plants along the streams and ditches are
brooklime, watercress, water starwort, creeping buttercup,
ivy-leaved crowfoot, marsh marigold, pondweed, and
forget-me-not. Ling, cross-leaved and fine-leaved heaths,
crowberry, purging flax, eyebright, gentian and milkwort
are found on the hills. Several rare plants occur in the
county. At the Culbin Sands the coralroot and the lesser
twayblade are found. The twayblade, the bird's-nest
orchis, the single flowered wintergreen, and the intermediate wintergreen occur in the Darnaway Woods; the
vernal squill and the Scottish lovage at Covesea. Smooth

gromwell may be seen between Burghead and Hopeman; sheep's-bit between Lossiemouth and Garmouth; bird's-foot in Urquhart; lesser skullcap on the Spey, near Rothes; bearberry near Lochindorb, and intermediate bladderwort in a moss below Grantown. Several species of alpine plants occur, as alpine lady's-mantle, alpine bedstraw, mountain sorrel, mountain speedwell, starry saxifrage, yellow saxifrage, and alpine enchanter's night-

The Duchess' Lime, Gordon Castle

shade. The last three are found on the banks of the Findhorn.

The soil of Morayshire is very suitable for Scots fir and larch; and these are the most extensively planted. Hardwood trees such as oak, beech and ash also thrive well. There is a magnificent lime tree called the "Duchess' Lime" at Gordon Castle. Its drooping branches cover an area the circumference of which

measures 300 feet. Fine beeches, oaks, elms, sycamores
and an old mulberry tree may be seen in the Cooper
Park, Elgin. The last of an avenue of venerable ash
trees stand by the side of the turnpike road entering
Forres from the west. In the garden of Burgie House is
a splendid sycamore, the girth of which is over 14 feet.

8. Weather and Climate.

The principal factors that determine the climate of a
country are its latitude, shape, altitude, exposure, ocean
currents, distance from the sea, mountain chains, winds,
character of its river systems and nature of its soils.

The weather of the British Isles is largely determined
by cyclonic disturbances which usually come from the
west or south-west. The movement of the air may be
cyclonic or anticyclonic. In a cyclone the pressure is
lowest in the centre and gradually increases outwards.
The wind blows in spiral curves around and also towards
the centre, the movement being counter-clockwise in the
northern hemisphere. The cyclone as a whole moves
forward, generally east or north-east in the latitude of our
islands. In the centre of the cyclone the air is rising, and
this air, which in our cyclones is moist, since they come
from the Atlantic, on rising to a higher (and therefore a
colder) level must produce rain. Hence cyclonic weather
may be regarded as rainy and cloudy weather. An anti-
cyclone is the opposite of a cyclone. The pressure is
greatest in the centre of the system and diminishes

outwards. The winds in this case blow spirally outward and their direction in our hemisphere is clockwise. Fresh supplies of air are received from above, and this air in descending gets warmer and accordingly no clouds are formed, clear skies indeed being a feature of anticyclonic weather. It is hot and calm in summer, with heavy dews at night ; in winter it is calm and cold. Only when anti-cyclonic conditions prevail do we in this country experience any considerable spells of fine weather.

All through the year the winds of Scotland are in-fluenced by three fairly permanent pressure centres, a high pressure area in the Atlantic about the latitude of the Azores ; a continental area in Europe extending into the west of Asia, high in winter and low in summer ; and a permanently low pressure area to the south of Iceland. In winter the continental and the Icelandic centres are predominant, and at that season south-west winds are most prevalent over our Islands.

Morayshire has long been famed for mildness and dryness of climate. The average temperature of Elgin throughout the year is about 49° F., and the average annual rainfall varies from 25 to 28 inches. Writing in 1640, Sir Robert Gordon of Straloch says, "in salubrity of climate Moray is not inferior to any....The air is so temperate that, when all around is bound up in the rigour of winter there are neither lasting snows nor such frosts as damage fruit or trees ; proving the truth of that boast of the natives, that they have forty days more of fine weather in every year than the neighbouring districts....While harvest has scarcely begun in surrounding districts,

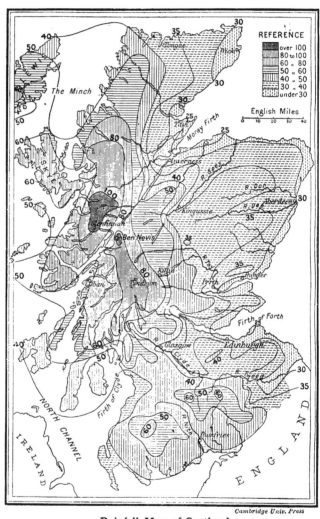

Rainfall Map of Scotland

(*By Andrew Watt, M.A.*)

there all is ripe and cut down...and, in comparison with other districts, winter is hardly felt."

The mildness of the climate is especially noteworthy when we consider that the coast of the county is within the 58th degree of north latitude. Along this southern shore of the Moray Firth lies the area of lowest rainfall in Scotland. The rain-bearing winds, being from the west and south-west, have been already deprived of most of their moisture in passing over the high hills before reaching the plain of Moray. This plain presents no object sufficiently high to cause the winds to part with the remainder of their moisture. Those westerly winds are föhn winds, i.e. winds whose temperature has been increased by the latent heat given out as the result of the condensation of water-vapour. This is undoubtedly the chief explanation of the mildness of the climate of this area. For the same reason falls of snow are less frequent and of less depth than in most parts of Scotland, and the snow continues for a shorter time on the ground than in the highlands in the south of the county. In the south the seasons are far less genial, and the climate is unpropitious and stormy. The spring is often very trying. In the more sheltered parts, however, the summer temperature is greater than on the coast-line, but the harvests are, as a rule, late and not infrequently precarious. The average annual rainfall of Grantown is fully 34 inches.

Most of the prevailing gales blow from the west or north-west. A south wind is comparatively rare and for fully 200 days every year the wind is from some part of the west. The most unpleasant weather is generally experienced towards the end of spring when frosty east winds often do great damage to the springing corn, the tender grass, and the blossoming buds.

9. The People—Race, Language, Population.

The earliest inhabitant of Moray seems to have been the man of the Neolithic Age, who ground and polished his implements of stone, made earthenware, and knew how to spin and weave. This race was conquered by the Gaelic branch of the Celts, who were followed to our country by another Celtic people, the Britons. In the second century A.D., Ptolemy gives the name Vacomagi to the tribe occupying part of the region between the Moray Firth and the river Tay. From the sixth century the inhabitants of the district north of the Grampians, to which Moray belongs, were called the Northern Picts.

Up till the end of the twelfth century Moray remained, for the most part, Celtic in population, institutions, and language. But from time to time bands of Norwegians settled on the seaboard.

With the reign of David I (1124–1153) settlers from England began to appear, and this went on steadily for several centuries. These Teutons succeeded in ousting the original race. To-day there are very few descendants of the Celts, who were once all-powerful in the county. It is probable that some descendants of the early Norwegian settlers are still to be found.

The dialect of Moray, as of Nairn, belongs to the North-East division of Scottish dialects, and as contrasted with the vernaculars of Banff and Aberdeen has practically no distinguishing peculiarities. One, however, which was referred to in the *Statistical Account*, still survives. This is the tendency to pronounce *v* as *w*. In this way *very* is pronounced *wery*; but this pronunciation is now used only by some of the oldest inhabitants. In the south of the county, to which the Celts were driven by Saxon intruders, Celtic place-names predominate, e.g. *Knockan-buie* (the yellow hill), *Allt Dearg* (the red burn).

The county is sparsely populated, the census of 1911 giving a return of 43,427 inhabitants. Two factors serve to complicate the comparison of this with the earlier censuses—one the alteration of the county boundaries, and the other the fact that the earlier census returns included as in Morayshire the populations of areas not belonging to the county, and included in the returns of other counties the populations of areas belonging to Morayshire. Adjusting the figures to suit the present area of the county, we find that the population in 1801 was 26,976. Only two intercensal decreases are reported since 1821, namely, a small decrease in 1891, and a considerable decrease, amounting to 1473, in 1911. The maximum population, 44,800, was recorded in 1901. The census of 1911 gives a population of 91 to the square mile. Lanarkshire, the most densely peopled county in Scotland, has 1633 to the square mile, while Sutherlandshire, our most thinly peopled county, has only 10. As compared with the census of 1901, four of the

six burghs in the county show a decrease.　Elgin shows
an increase of 3·0 per cent. and Lossiemouth an increase
of 7·8 per cent.　The increase in Elgin is only what
might be expected when we consider that, being one of
the most attractive towns in Scotland, it is becoming
more and more popular as a residential town and as a
pleasure resort.　The large increase in Lossiemouth is

Lossiemouth golf-links

due almost entirely to its excellent golf-links.　Indeed
Lossiemouth forms a typical example of what we may
call a "golf town."　Not only is its golf-course one of
the finest in Britain, but here also the visitor finds a
highly bracing climate, while to the antiquarian, the
geologist, the botanist, and the artist, the town is an

ideal centre. The apparent increase in the population of Forres in 1911 as compared with 1901 is due to an alteration of the burgh boundaries.

Of the nineteen parishes in the county only four have a larger population than in 1901. These are Birnie (increase of seven), Drainie (containing Lossiemouth), Elgin (containing the burgh), and Alves. The largest decreases are reported from Rothes, Cromdale, Forres, Knockando and Urquhart.

The outstanding feature in the recent census returns of the county is the uniform decrease in the agricultural population. This is largely due to migration to manufacturing centres, to emigration, and to the cheapness of imported food supplies. Another important factor, however, is the introduction of labour-saving agricultural machinery.

Agriculture gives employment to 3506 males or 27·5 per cent. of the total occupied males, and of these 942 are farmers or crofters. There are 1285 fishermen or crofter-fishermen in the county. Of the female population 2091 are employed in domestic service, while 406 are dressmakers, 223 teachers, 180 laundry-workers, and 160 commercial clerks. No fewer than 153 are farmers or crofters, while 291 more make a living at farm work.

Those returned as speakers of Gaelic number 1214, and all of them, with one exception, can speak English. Only 287 of the Gaelic speakers were born in the county. The rest belong to Inverness, Ross and Cromarty, or Sutherland, while five were born in Ireland.

10. Agriculture.

It was not until the eighteenth century that anything worthy of the name of agriculture was carried on in Morayshire or indeed in any part of Scotland. Up to that time Morayshire was poor, large tracts of the county were unenclosed, and only the driest parts were cultivated. And yet such were the mildness and dryness of the climate and the excellence of the soil that no other part of Scotland has had so many spontaneous testimonials borne to its fertility. Whitelock, for example, about the middle of the seventeenth century, says, "Ashfield's regiment was marched into Murray-land, which is the most fruitful country in Scotland." With the advent of various improvements in farming—such as, cropping by rotation, the introduction of root crops, the sowing of grass and clover, the carrying out of drainage schemes, the application of lime and artificial manures to the land —the coastal plain of Moray became known as "the Granary." Thereafter attention was devoted to the careful breeding of stock.

Agriculturally, as geologically, the county of Moray falls into two well-marked divisions—the northern coastal plain, or the "Laigh of Moray," and the highland district in the south. For arable purposes the coastal plain is all-important, being favoured with a rich loamy soil, and great mildness and dryness of climate. In the uplands of the county the land is poorer, while the climate is cold and damp. The inhabitants, accordingly, devote their attention mainly to pastoral pursuits.

Harvest time, Elgin

It would be almost impossible to exaggerate the importance of the Morayshire Farmers' Club to the agriculture of the county. Founded in a modest way as far back as 1799, the club is still vigorous and progressive. Sheriff Rampini thus describes the institution : "Whatever else it was not, the club was certainly practical. It introduced new implements of agriculture ; it bought studhorses for the use of the district ; it instituted shows; it gave premiums for excellence in almost every department of agricultural life. What was of even more importance was, that it strove to instil a hopeful spirit into the agricultural community." The proud position which Morayshire holds among the agricultural counties of Scotland is due in no small measure to this enlightened club.

Of the 304,931 acres of land in the county, 98,939 were under crops and grass in 1912. Of these, 91,058 were arable while the remaining 7881 acres were permanent grass. 46,192 acres were under woods and plantations. Nearly one-third of the whole area is under crops. There were in all 1715 holdings for agricultural purposes. Of these, 345 were under five acres, 754 from five to 50 acres, 578 from 50 to 300 acres, and 38 above 300 acres. In 1912 the acreage under the principal crops was as follows :

Oats ...	...	...	...	...	23,588 acres
Barley ...	...	...	...	...	10,754 ,,
Wheat...	...	...	...	...	677 ,,
Rye ...	...	...	...	...	527 ,,
Potatoes	...	...	...	...	1971 ,,
Turnips	...	...	...	...	14,750 ,,
Clover, Grasses, etc. (in rotation) ...					38,383 ,,

The large quantities of oats, turnips and clover are required for the live-stock industry, while the numerous distilleries in the county explain the relatively large amount of barley grown.

Agriculture is pursued in a spirited manner, and as a general rule land yields a higher rent than in the neighbouring counties. Wherever possible, practically all the

Aberdeen-Angus bull

former waste land is now under cultivation or planted with trees. In the lower part of the shire the proprietors, and notably the Earl of Fife, made plantations to such an extent that almost every part of the land unsuitable for tilling is covered with trees. The plantations mainly consist of larch and fir. Much timber was at one time floated down the Spey and other rivers; and with the increased facilities for carriage offered by the railways wood-cutting was carried on to a still greater extent. Oats is the chief crop in the south of the county, while the north is unsurpassed for the quantity and the excellence of its barley, turnips, oats, and potatoes.

But it is more in stock-rearing than in the raising of crops that the Morayshire farmer has distinguished himself. The kinds of cattle reared are Aberdeen-Angus, shorthorns, crosses of these breeds, and a few Highland.

Morayshire forms part of the great beef-producing district of Scotland, the district in which the Aberdeen-Angus cattle have their home. The cows are fair milkers, but the cattle are most famous as beef-producers. It is largely, if not entirely, owing to them that Scotland is assigned the premier place in the great English beef markets. "Prime Scots," as the beef is termed, almost invariably tops the quotations, and the bulk of the consignments proceeds from the north-east corner of Scotland, of which Morayshire forms part. No class of cattle takes on beef so rapidly and matures so early as the Aberdeen-Angus. Hence the Morayshire farmers are becoming yearly more extensive stock-raisers. The breeders in the county have shown great shrewdness and skill in the management of their cattle, and in no part of Scotland are there finer herds of those beautiful black polls.

Shorthorns are also favourites, being valued for their milking qualities, and still more as beef-producers. They are extensively used for crossing with the Aberdeen-Angus, a very satisfactory cross.

In the hilly district in the south of the county

Highland cattle are reared.   Owing to their sturdy constitutions they seem proof against wind and weather. Their sustenance is largely picked up from the natural herbage of their mountain home. The best cross procured from this breed is with the shorthorn. The resulting breed possesses the most valuable characteristics of the Highlander with the rich milking and early-maturing qualities of the shorthorn.

It is interesting to note that there are fewer cattle bred in Morayshire to-day than, say, twenty years ago. This is because many of the farmers have stopped the breeding of cattle altogether, as they find it more lucrative to devote their attention to feeding.   Large numbers of cattle are bred in the uplands of the county and sold as calves or yearlings to the farmers of the Laigh.   Others are brought from the north of England and from Ireland to be sold off when they are from two to two and a half years old.   The total number of cattle in the county in 1912 was 22,331.

Blackfaced, Leicester, and Cheviot sheep are reared, but most of the sheep-stock consists of crosses of these breeds.   The crosses feed very fast and also produce the best of mutton.   The low-lying parts of the county are well adapted for the feeding of sheep, as the soil is naturally dry and the crops of good quality.   The mountain breeds are the Cheviots and the blackfaced ; and hence we find these in the highlands, but the crosses are the favourites in the rich lowlands.   Even in the highlands, however, the Cheviots are being displaced by the hardier blackfaces, as the latter are much better suited to

the trying conditions of the southern parts of the county.
The lambs of the high-lying areas are bought by the
farmers of the northern plain, where they are fed and
sold as yearlings. By dint of careful and judicious
selection the blackfaced sheep have attained a high
degree of development. They thrive remarkably well
under most trying conditions, and are the only breed of
sheep that can be safely trusted to live through a severe
winter in the uplands of the county.

The Leicesters take on fat with great rapidity but
their mutton is coarse. Their chief value lies in the
adaptability of the Leicester ram for crossing purposes,
and the cross between the Leicester ram and the Cheviot
ewe is the principal sheep-stock on the lowland arable
farms. This cross is perhaps the most profitable stock
that can be reared. In 1912 the number of sheep in the
county was 53,754. In the same year the number of
horses was 5052, and of pigs 2426.

11. Industries and Manufactures.

Fishing ranks next to agriculture among the occupa-
tions of the people. The chief fishing ground is the
Moray Firth. Drift-nets and lines are used. Herrings,
haddocks, cod, ling, and flounders are caught. The most
important fishing centres are Lossiemouth, Hopeman, and
Burghead. The fishing carried on at Findhorn is incon-
siderable. In 1911, 1408 males and 193 females were
engaged in the fishing industry. The salmon fisheries

along the coast and on the Spey and the Findhorn are of great value. At the Fochabers hatchery tens of thousands of salmon ova are hatched every year.

Distilling is an important industry. There are about a score of distilleries in the county. The very finest malt whisky is produced. The rise of the industry may be attributed partly to the excellence of the barley grown in the Laigh and partly to the suitability of the water for distilling. Draff, a distillery by-product, is largely used for feeding cattle, while another by-product, maltassa, is a valuable manure. Elgin manufactures woollens, tweeds, plaidings, and leather, and has also breweries, saw-mills, corn-mills, nurseries, and an iron-foundry. At Burghead and Forres are chemical works.

The greatest drawback is the scanty supply of minerals. Owing to the absence of coal what little mineral wealth there is cannot be profitably worked. At Lossiemouth lead ore has several times been worked, always with financial loss. The minerals raised in the county in 1909 were :

Brick clay ...	...	...	6750	tons
Gravel and Sand	...	...	7968	,,
Igneous Rocks	...	...	6518	,,
Sandstone ...	...	...	23,688	,,

Sandstone quarries are numerous, the stone showing great variety of texture and colour. Practically all the houses in the county are built of local stone.

The shipping of Morayshire, carried on at Lossiemouth (the port of Elgin), Burghead, Hopeman, and Findhorn, is comparatively unimportant. Garmouth is an extinct port. The *Statistical Account* tells us that a century ago, " the quantity of natural fir timber sold here [i.e. at Garmouth] exceeded £40,000 Sterling yearly, and even so recently as 1818 it amounted to about £30,000." The timber was brought from the forests of Abernethy, Glen More, and Rothiemurchus. The Salmon Fishing Company employed at Garmouth twelve crews of seven

Burghead

men each. In 1818 no fewer than 257 vessels entered Garmouth harbour. The floods of 1829 left the port dangerous to shipping.

Kingston (so called after a firm from Kingston-upon-Hull who made extensive purchases of wood in Glen More) at one time did a considerable trade, and its decay is also due to the vagaries of the Spey. In 1834 over

200 vessels called at this port, which was also a well-known ship-building centre.   The occasional launching of a small fishing-boat is all that remains now to remind us of the one-time flourishing industry of Kingston.

12. History.

The history of Morayshire is inseparably bound up with the history of the Province of Moray.   As far back as authentic records go, this province was one of the important territorial divisions of the country, and such indeed it remained till the twelfth century.

It is uncertain whether Severus during his great campaign in 208 reached the Moray Firth, but echoes of his march and of the guerilla attacks upon him would undoubtedly have reached the region.   In later centuries Moray formed part of the dominion of the Northern Picts.   Their conversion to Christianity was effected under the auspices of Columba towards the end of the sixth century, and marks an important stage in the history of the county.   At this time Burghead was selected as the site of a Christian church.   The new religion flourished for a time, but in 717 the Columban Church was expelled by King Nectan.   This was the end of the Irish Church in the province, for, when next we read of Christianity as a force in these parts, it was the Roman form.   The history of our country in the first half of the ninth century is very obscure, but in 844 Moray became nominally part of the dominions of Kenneth MacAlpin when he united the kingdom of the Picts to the kingdom of the Scots.

The peace of the country was next broken by the Norsemen, who came "plundering, tearing and killing not only sheep and oxen, but priests and Levites, and choirs of monks and nuns." Their onslaughts struck terror into the hearts of the inhabitants of the province. The first Norse conqueror of Moray was Thorstein the Red (874). About ten years later Sigurd "conquered... Maerhaefui [Moray] and Ross, and built a borg [probably Burghead] on the southern borders of Maerhaefui." Before very long, however, the mormaers of Moray managed to throw off the yoke of the Norsemen—but only for a time. In the eleventh century Moray stretched along the seashore from the mouth of the Spey to the Dornoch Firth with the hinterland to the source of the Spey and to the watershed from about Glen Quoich to Loch Broom. On 14 August, 1040, the most important battle ever fought in Morayshire took place. The formidable Scots army was commanded by King Duncan. The men of Moray were led by Macbeth, their mormaer, one of King Duncan's most distinguished generals. Thorfinn, earl of Orkney and Shetland, commanded the Norsemen, and completely routed the Scots. King Duncan, according to tradition, sought rest at Bothgowan (the smith's bothy), now Pitgaveny, where he was slain by Macbeth. Macbeth then seized the throne, to which he had, through his wife Gruoch, a slight claim, and made an able ruler. The Norsemen gave no more trouble; Thorfinn became Macbeth's ally. Seventeen years later

Macbeth was defeated and slain by Duncan's son, Malcolm Canmore. The great tragedy of Shakespeare is based on Holinshed, who reproduced the perverted story of the Scottish historians.

Moray continued to be ruled by its mormaers (or earls, as they were hereafter called), some of whom acted like independent princes. This state of matters continued

Duncan's bedroom, Cawdor Castle

till the reign of David I. At the battle of Stracathro, some three miles north-east of Brechin, the Earl of Moray was defeated and slain. Five years later the subjugation was completed with the aid of barons from the north of England—including several Normans—among whom David apportioned the lands of Moray. The earldom was then made definitely dependent on the crown, whose authority David tried to strengthen by the foundation of a monastery and the erection of several castles. Only three years later we find the men of Moray fighting under David's banner at the Battle of the Standard.

But Moray was still far from enjoying the blessings of uninterrupted peace. The clansmen and their chiefs made frequent raids from their highland fastnesses on the plain of Moray. In this way the inhabitants of the Laigh were kept for generations in a state of terror, until ultimately the civilizing influence of the church made itself felt.

The struggle for independence under Wallace and Bruce also tended to make the men of Moray unite, lowlanders and highlanders alike. In 1296 and in 1303 Edward I came as far north as Elgin. On the first occasion he was welcomed with every sign of cordiality. The burgesses of the town, the bishop and clergy of the diocese, and many knights and gentlemen acknowledged him as their king. So promising did everything appear that Edward, before his four days' sojourn in "la cité d'eign" was over, issued summonses for the memorable Parliament of Berwick.

Wallace, however, had no more loyal supporter than Sir Andrew Moray, under whose leadership resistance to the English spread through the district like wildfire. The royal castles—including those of Forres and Elgin— were attacked and burned; and Andrew Moray may well be regarded as Wallace's right hand in the great uprising. Unfortunately, however, Moray met his death at the battle of Stirling Bridge (1297). In 1303 Edward,

at the head of a mighty army, again entered Scotland,
devastating and pillaging wherever he went. From
Aberdeen he marched by Banff and Elgin to Kinloss.
Then he turned southward and without great difficulty
reduced the castle of Lochindorb, the stronghold of the
Comyns, where he stayed for a month in order to receive
the submission of the chiefs. In the events leading up
to the triumph at Bannockburn Morayland played a
conspicuous part. Evidence of this is found in the fact
that Bruce immediately after that victory raised the
Province of Moray to the dignity of an earldom for
his nephew, Thomas Randolph. Bruce granted him a
charter of a very unusual kind, whereby the free tenants
and the barons of the earldom, who held of the crown in
chief, should henceforth hold of the earl. This curious
relation towards the crown was long held by the burgh
of Elgin. Few earldoms shine so brightly in the annals
of our country. Randolph, the first earl, " Black Agnes
of Dunbar," "the Good Earl of Moray," are names
familiar to every student of Scottish history.

On 3rd July, 1650, Charles II landed at Garmouth
after his first exile in Holland. This campaign was to
end at Worcester. The battle of Cromdale (1690),
where General Buchan was defeated by the Government
troops under General Livingstone, was the expiring flame
of the Jacobite rising, which had been ruined by the
death of Dundee at Killiecrankie, and by the disaster at
Dunkeld. In 1715 and in 1745 practically all Morayshire was solid in its allegiance to the Hanoverians.
A few weeks before Culloden, Prince Charles Edward

marched into Moray, where he spent eleven days, mostly
in Elgin.   He stayed in Thunderton House, the occupier
at that time being a Mrs Anderson, an ardent supporter
of the Jacobite cause.   About a month later the Duke of
Cumberland passed through the town in great haste, and

Thunderton House

shortly afterwards shattered Jacobitism on the bloody
field of Drummossie (1746).

Morayshire has twice suffered devastations without
parallel in the history of Scotland.   The first of these
was the destruction of the Barony of Culbin, already
mentioned.   The other calamity is well known as the

"Moray Floods" of 2nd, 3rd and 4th August, 1829, splendidly described by Sir Thomas Dick Lauder. "The noise," says he, "was a distinct combination of two kinds of sound—one a uniformly continued roar, the other like rapidly repeated discharges of many cannons at once.... Above all this was heard the fiend-like shriek of the wind, yelling as if the demon of desolation had been riding upon its blast....The rain was descending in sheets, not in drops, and there was a peculiar and indescribable lurid or rather bronze-like hue, that pervaded the whole face of nature as if poison had been abroad in the air." At one place on the Findhorn observations showed that the river had risen 50 feet above the usual level, while the Divie, its tributary, rose 40 feet. In the plain of Forres the Findhorn covered some twenty square miles. All the low-lying ground round Elgin was inundated by the Lossie. Bridges were swept away on every hand. Dwelling-houses were demolished. Arable land and forest were carried off to the sea by the acre. Crofters watched their homesteads as they sailed towards the Firth. The loss due to the floods was enormous. The bridge across the Spey at Fochabers, which was destroyed, cost £14,000 only twenty-five years previously. The Duke of Gordon's loss amounted to over £16,000. But it was the poorest classes that suffered most, for in many cases they lost everything—even their few hard-earned and well-saved banknotes were whirled off to the ocean.

13. Antiquities.

Morayshire is rich in relics of prehistoric times. Numerous flint implements of the Neolithic Age have been found. Thousands of chipped flints have been picked up at the Culbin Sands, which have, therefore, been called "The Armoury." Perhaps here was a factory from which the surrounding districts were supplied with weapons of the chase and of war, and simple implements for domestic use. Culbin flints show considerable skill and taste in workmanship. Another probable manufactory was discovered in the parish of Urquhart in 1870.

Stone circles are numerous. Unfortunately many of these interesting memorials have been cleared away before the advance of the plough. Others have been dismantled, and in many cases some of the boulders have been removed. One of the best preserved is on the farm of Viewfield in the parish of Urquhart. Another example is at Cowiemuir in the parish of Bellie. The fragments of a third circle are found in Knockando.

In many of the circles urns have been found, thus showing that the builders, if they also deposited the urns, knew something of the potter's art and practised cremation. The circles further show that the builders were familiar with concentric circles, and were able to take measurements with considerable exactitude. One peculiarity is that the highest pillars are always found to the south. The outer circle of stones shows noon and also the solstices and equinoxes. From this it is argued that the circles, in addition to serving as tombs and representing some religious idea, did duty as clocks and calendars. The circles are of great antiquity. From the bronze articles found in them, the style of burial, and the nature of the pottery associated with them, we may regard them as belonging to the Bronze Age.

Bronze armlet, from the
Culbin Sands

Cist graves have been discovered in various parts of the county, for example, on the estate of Inverugie in Duffus, and in the parish of Urquhart. Such graves usually contain an urn. One grave on the farm of Meft in Urquhart yielded two urns full of ashes and burnt bones. In the same parish, some fine gold armlets have been found. On Burgie Lodge Farm, a cist containing the skeleton of a brachycephalous man was unearthed in 1913.

Sweno's Stone, which stands close to Forres, is perhaps the finest sculptured stone in Scotland. It represents the high-water mark of Celtic art. "On the south side," says Rhind's *Sketches and Antiquities of Moray*, "there are five divisions, each filled up by numerous figures cut in relief. The first division represents a number of persons as if engaged in deep council, and holding conversation in groups....The second division exhibits an army of

horse and foot on the march, the cavalry being in the
van and at full gallop, the infantry following with

Sweno's Stone, Forres

spears in their hands and shields. In the third division
are appearances of a battle....The fourth division shows

a number of captives bound together,...while a row of
warriors above, with unsheathed swords, are shouting
victory....The other, or north side of the stone, has
only three divisions. Below are two figures, with human
heads, though their bodies are of rather grotesque forms,
typical perhaps of priests, bending over something as if
in an attitude of prayer....In the division above is a long
cross, the arms at the top being within a circle." What
particular event, if any, the stone was erected to commemorate remains, and is likely to remain, a mystery.

The "Roman Well" or bath at Burghead is a cistern
hewn out of the solid rock. Its four sides are fully 10 feet
in height, and the basin is rather more than 4 feet in
depth. Steps lead down to the basin. As it has never
been proved that the Romans gained a permanent footing
on the shores of the Moray Firth, it is more than probable
that the well is not Roman. And whether it represents
an erection for baptism by the early Christian church is a
highly disputable point.

The two unequal terraces at the extremity of the
peninsula on which
Burghead stands were
defended by ramparts
and earthworks, but few
traces of these now re-
main. The probability
is that on this promon-
tory was a Pictish tower
or broch, and that it fell
into the hands of the

Bull stone, from Burghead

Norsemen, who enlarged it and strengthened it. Numerous objects of archaeological interest have been discovered on the promontory. Amongst these are several sculptured bulls, a jug composed of a mixture of tin and copper said to have been found in the "Roman Well," and a silver mounting for a horn.

14. Architecture—(*a*) Ecclesiastical.

The first buildings in Scotland erected for purposes of worship that we can call churches were those built by Columba. Consisting as they did of one small rectangular chamber with a single door and a single window, they do not possess any architectural qualities.

The Norman style of architecture in Scotland dates from the twelfth century. It is characterized by "cubical" masonry, semicircular arches, and quaint carving, especially on the doorways (the arches of which are moulded) and chancel arches. This style is represented in Morayshire by the Church of Birnie.

In the thirteenth century the semicircular Gothic arch disappeared in favour of the pointed arch—the First Pointed or Early English style. This style was characterized by long narrow or "lancet" windows, which gave an impression of greater lightness and spaciousness. The gables and roofs were high, and the pinnacles and spires simple in the extreme. In Scotland the Gothic arch was retained, and in this way the Scottish style was less pure than that of England or France. The earlier parts of Elgin Cathedral supply excellent examples of this type.

From about 1274 to 1377 the Second Pointed or Decorated style was prevalent in Scotland. The windows were divided into several lights, and their upper parts were filled with fine tracery. This may be regarded as the most perfect and beautiful style of Gothic architecture and is seen in the south aisle windows of Elgin Cathedral.

The Kirk of Birnie

The stiffer Third Pointed or Perpendicular style prevailed from the end of the fourteenth to the middle of the fifteenth century. There is an upright and square tendency in the tracery of the windows, which were further characterized by perpendicular lines.

The Church of Birnie is evidently ancient; and the

square-dressed freestone ashlar, employed externally and internally, at once marks out the building as of Norman workmanship. With the exception of the chancel arch, every part of the building is perfectly plain. Within the chancel arch is a stone font of Norman design. The church also contains an extremely rare and highly interesting specimen of the square-shaped Celtic bell. It is conjectured that the church may have been preceded by a Celtic monastery.

Kinloss Abbey is one of the monastic establishments founded in Moray by David I, and is an example of the Transition style intervening between the First and the Second Pointed. According to legend, the king had lost his way in the woods while hunting, and was guided by a deer to an open space where the Virgin appeared and ordered him to build there a church to her honour in recognition of his deliverance. The abbey, founded in 1150, was colonized by Cistercian monks from Melrose. David endowed the abbey with lands, and several of his successors made munificent grants for its upkeep and extension. About 1650 the materials of the abbey were taken to Inverness and used in the construction of Cromwell's citadel. Accordingly the ruins of Kinloss Abbey are the merest fragments. Very little ornamentation remains. The gateway is clearly transitional, though several of the details are Norman.

In Elgin Cathedral the county possesses one of the noblest examples of Gothic style in the kingdom. Indeed, it is generally admitted that the fragments of this glorious edifice rank with the finest examples of mediaeval architecture.   Its ruins are "among the most melancholy and the most impressive in Scotland."

The See of Moray was founded by Alexander I early in the twelfth century.   The seat of the bishop, however, varied from time to time, till in 1203 Pope Innocent III decided that it should be at Spynie.   This situation being found inconvenient, Pope Honorius changed it to the banks of the Lossie, where Alexander II, a great admirer and patron of Elgin, granted a fine site for the erection of a cathedral, which Andrew, Bishop of Moray, started building in 1224.   The cathedral was dedicated to the Holy Trinity.   By 1270 the building must have been well advanced, for Fordun relates that in that year the cathedral and the canons' dwellings were destroyed by fire.   In 1390 the cathedral suffered at the hands of Alexander Stewart, son of King Robert II, but better known as the "Wolf of Badenoch."   In revenge on the bishop who brought about his excommunication, the "Wolf," at the head of a strong horde of retainers, swept down upon Elgin and left "the lantern of the North," as the cathedral was called, a blackened mass of ruins.   Luckily the catastrophe, terrible though it was, only partially destroyed the building.   The king ordered the "Wolf" to do penance and to assist in rebuilding the cathedral.   The work of restoration proceeded slowly and probably lasted during most of the fifteenth century. The interior of the chapter house, and other parts of the building, are undoubted examples of fifteenth-century architecture.

By the middle of the sixteenth century the period of

decline set in. In 1568 the Regent Moray removed the lead from the roof in order to provide money for his soldiers. The rafters were blown down by a storm in 1637 and the destruction of the interior followed soon after. In 1640 the painted screen between the nave and the choir was broken down by order of the General Assembly. The tracery of the great west window was next demolished, an act of vandalism attributed to Cromwell's troops during the Commonwealth. But the principal disaster was yet to come—the falling of the central tower on Easter Sunday, 1711. This brought about the complete destruction of the nave and transepts. For a whole century thereafter the ruins served as a quarry.

In beauty of architecture and in completeness, Elgin Cathedral must be placed in the front of our Scottish ecclesiastical buildings. It had a large nave and double aisles, a choir and presbytery, north and south transepts, a lady chapel, an octagonal chapter-house, a very high tower and spire over the crossing, two splendid towers at the west end, and two beautiful turrets at the east end. Most of the existing parts were erected during the thirteenth century—the heyday of Scottish ecclesiastical architecture. The fine fragments of sculpture and decoration still to be seen in the chapter-house are sufficient evidence of the beauty and excellence of the internal work. In some respects, notably that of the western doorway and the pointed windows of the choir, Elgin Cathedral is unsurpassed, perhaps unequalled, by any other cathedral in Britain. In dimensions, the building is somewhat inferior to the cathedrals of Glasgow and

The Western Doorway, Elgin Cathedral

St Andrews. The central aisle measures 118 feet, and the total length of the nave is 84 feet. The present chapter-house belongs almost entirely to the Third Pointed period. There is no other example in Scotland of a detached octagonal building with vaulted roof and central pillar. Opposite the entrance are fine arches containing Third Pointed ornaments.

Pluscarden Priory, about six miles south-west of Elgin, dates back to the thirteenth century. In 1398 the buildings were in a sorry state of disrepair; and nearly a century later, this priory and Urquhart Priory (established in 1125 for Benedictines from Dunfermline) were united. Curiously enough, the monks were left in undisturbed possession at the Reformation. At present there may be seen north and south transepts, an aisleless choir, a square tower, a sacristy called St Mary's Aisle, a chapter-house, and a monks' hall. In the transepts with their eastern aisles the First Pointed style is evident, and the buildings therefore date back to the thirteenth century. The same style is seen in the sacristy. Architecturally the chapter-house is specially interesting. The apartment is nearly 30 feet square and has a central pillar. The door and windows show First Pointed work (thirteenth century), while the central pillar is unmistakably of fifteenth-century workmanship.

The ruins of Altyre Church also belong to the First Pointed period. The structure, characterized by extreme plainness, consisted of one rectangular chamber. The side windows are simple lancet; the east wall has a window with branched mullion.

Greyfriars Church, Elgin, was erected about 1479. Belonging to the Franciscan order of mendicant friars, it is extremely plain and simple. After the Reformation, the building ceased to be used for service. Montrose plundered the house, but fortunately left the church unharmed. The ruins have been restored, thanks to the efforts of the late Marquis of Bute.

Michael Kirk, situated half a mile west of Drainie,

Pluscarden Priory

was erected in 1703 as a mausoleum by Lodvic Gordon, a proprietor in the neighbourhood. In the east and west gables are large lancet windows with beautiful tracery. This tracery, though slightly suggestive of Gothic, is inclined to the Classical. Some of the ornamentation is rather curious, if not inappropriate. Thus in the east window may be seen a series of Cupid's heads. The Renaissance influence is noticeable in the urns on the gables. This survival of Gothic so far north has a historical as well as an architectural interest, for the Episcopal form of worship found very considerable support in this part of Scotland.

Of modern churches, the Parish Church in the High Street of Elgin should be mentioned. Its classical frontage is surmounted by a tower 112 feet in height. St Columba's Church, Elgin, is a fine example of recent developments in the ecclesiastical architecture of Scotland.

15. Architecture—(*b*) Military.

Fortifications built of stone and lime were probably first introduced into Scotland by the Normans in the eleventh or the twelfth century. They were quadrangular buildings, called keeps, and most of the Scottish strongholds were modelled on them. Duffus Castle is an example of a simple keep. It stands on a mound in the centre of a plain, and is surrounded by a fosse enclosing an area of about nine acres.

Darnaway Castle, once the seat of the Earls of Moray, stands on an eminence in the midst of a large forest. The more ancient parts date back to the middle of the fifteenth century. The ancient oaken roof of the hall has been preserved. It is clearly of fifteenth-century workmanship, and, with the single exception of the roof of the Parliament House in Edinburgh, is the only example of this style in the country. The front is modern, dating from 1810.

Spynie Palace is an interesting ruin. It was originally situated on Loch Spynie but, owing to the draining of the loch, a considerable stretch of land now lies between (see p. 21). The site was selected at the time when Spynie was the seat of the Cathedral of Moray. Designed as a building surrounding a courtyard, it was one of our finest examples of fifteenth-century castellated architecture. The

Old Duffus Castle, Elgin

outside walls remain, but the interior of the building is a complete ruin.

Coxton Tower, on a gentle slope two miles east of Elgin, is one of the most remarkable keeps in Scotland. Over the doorway is the date 1644, but some authorities think that the tower is much older. Though there are

Coxton Tower

numerous shot-holes, the building is far from strong as a defence.

The Bishop's House, Elgin, is a fine example of six-teenth-century architecture. Its most noteworthy external feature was the small oriel window in the east front, which has recently fallen.

In Innes House, five miles north-east of Elgin, the keep plan is departed from and the more commodious L plan is introduced. The building dates back to the middle of the seventeenth century, the architect being the famous William Aitoun. The chief architectural interest of Innes House lies in the fact that here, on the border of the Moray Firth, Renaissance ornamentation is found on a building designed on the old L plan.

Z plans, or plans with two towers at diagonally op-posite corners, were adopted after the introduction of fire-arms, the two wings making it possible for the defenders to guard the approaches to the castle on all sides. This style is exemplified in Burgie Castle, about two miles south of Kinloss station. The greater part of the ancient building has been removed. A six-storeyed square tower still remains. The battlements and their ornamentation are well preserved.

Blervie Castle, of which now only a fragment remains, occupied a fine position about a mile north-east of Rafford, on a hill commanding the pass by which the Highland Railway runs to the south of the county. We learn from the Exchequer Rolls that this castle was specially repaired and garrisoned before the invasion of King Haco.

16. Architecture — (c) Municipal and Domestic.

For a provincial town, Elgin has several remarkably good examples of municipal architecture. The Town Hall, in Moray Street, is a handsome building. It has an imposing pinnacled tower crowned with a cupola. The Victoria School of Science and Art is only a few yards distant from the Town Hall. Opposite the Victoria School, and standing in its own grounds, is the Elgin Academy. The building, though plain externally, is substantial and commodious. Gray's Hospital, situated on a beautiful site at the west end of High Street, at once commands attention. The foundation was laid in 1829, the funds having been bequeathed by Dr Alexander Gray, a native of the town. Mr James Shepherd, also a native of Elgin, gave £10,000 for the purpose of equipping the hospital on modern lines. Anderson's Institution, a two-storeyed quadrangular edifice with a handsome dome, is named after the founder, Major-General Anderson, whose object was "the education of youth and the support of old age." For this purpose he left the sum of £70,000. The building was opened in 1832.

The fact that in the thirteenth century Elgin was the See of the Bishop of Moray tended to improve the style of its buildings. The town, however, has lost much of the highest architectural interest through the nineteenth-century craze for modernizing. A few interesting features still remain, among them a fine angle-staircase tower, and one or two piazzaed houses, i.e., houses whose front walls rest on arcades.

An old house at Elgin

Domestic architecture falls into two classes—ancient fortalices, to which additions have been made for the purpose of bringing them up to date as mansion-houses, and dwelling-houses of recent construction. To the first class belongs Easter Elchies, an old building situated on the north bank of the Spey, about a mile from Craigellachie. It belonged to Lord Elchies (1690–1754), but it was completely reconstructed in 1857. Two miles farther up the river is Wester Elchies, a fine example of an ancient Scottish mansion, to which a castellated structure has been recently added. Castle Grant is situated near the extreme south of the county and is a massive pile. Originally designed on the L-plan, the building was completely surrounded by new buildings by Sir Ludovick Grant (1743–1773). Brodie Castle, three miles west of Forres, is a good example of an ancient stronghold modernized in recent times. For over 500 years the castle has belonged to the Brodies of Brodie. Gordon Castle, Fochabers, a vast quadrangular Gothic pile built on what was formerly a morass called the Bog of Gight, is the seat of the Duke of Richmond and Gordon.

17. Communications.

For the purpose of connecting the Northern Lowlands, of which Morayshire forms part, with the more important Central Lowlands, there have been at all times two main lines of communication. The more direct runs through the heart of the Grampians. It ascends from the Tay by the valleys of the Tummel and the Garry to the Pass of Drumochter, a height of 1500 feet above sea-level. Then taking advantage of Glen Truim, it descends through Badenoch and reaches the valley of the Spey. The other route passes to the east of the Grampians. It goes up the valley of the Ury and finally reaches the lowlands by the valley of the Deveron.

Another route connects Strathmore with the valley of the Dee. Afterwards it turns northwards and then north-westwards across Banffshire to the Spey. Thereafter it follows the Glen of Rothes to Elgin. This route is of historical interest, having been followed by Edward I of England on his return journey from Elgin in 1296. On his northward journey Edward arrived at Banff by the Glen Ury route. The close connection between the main routes of the district and the physical features is evident.

To-day the main roads and railways follow the routes we have described. On the whole, the railway gradients are easy, and the construction of railways did not involve very serious difficulties. The Inverness and Keith section of the Highland Railway enters the shire at the Spey, and passes through it from east to west by Lhanbryde, Elgin, and Forres, whence it proceeds to Nairn. A branch line runs from Alves station to Burghead and Hopeman. At Forres, the Forres and Perth section branches off and passes through the county from north to south, leaving it about four miles south of Grantown-on-Spey, almost at the most southerly point of the shire. Starting from Elgin as its northern terminus, the Great North of Scotland Railway system has a branch line to Lossiemouth. Its construction over the coastal plain was a simple matter, and the railway runs for most of its length in a perfectly

straight line. The main line passes southwards through
the Glen of Rothes, adopting the very route by which
Edward I's troops marched homewards. After passing
Rothes the line takes advantage of the valley of the Spey
and leaves the county on crossing that river at Craigellachie. At Craigellachie the line branches, one part
passing to Aberdeen *via* Keith, the other turning up
Strathspey. For the first six miles the Strathspey section

Grantown-on-Spey

runs on the Banffshire side of the river, but at Carron it
crosses to Morayshire, where, with the exception of about
three-quarters of a mile near Ballindalloch, it remains till
it passes into Inverness-shire, about two miles east of
Grantown. A branch line runs along the coast from Elgin
by Buckie, Cullen and Portsoy to Tillynaught Junction.
In 1858 railway communication was established between
Rothes and Orton, but after the adoption of the present

more direct route from Rothes to Elgin, *viâ* the Glen of Rothes, this branch was discontinued.

Lossiemouth and Burghead have regular steam communication with Aberdeen and Leith.

18. Administration.

In early times the office of sheriff became hereditary and was held by one of the leading nobles of the county. The first heritable Sheriff of Morayshire whose name is found in the records is Alexander Douglas (1226), but there must have been several before him. At first the sheriffdom of Moray extended far beyond the present boundaries of the county. As the result of the '45 rebellion heritable jurisdictions were abolished. Morayshire has a Sheriff-Principal, who is also Sheriff of Inverness-shire and Nairnshire.

In 1782 a Lord-Lieutenant was appointed for each county. He is selected from one of the county families and is assisted by a number of Deputy-Lieutenants and Justices of the Peace. The Lord-Lieutenant is the king's representative at any important county function.

The chief administrative work falls upon the County Council. This body was first established in 1889, and the Chairman is called the County Convener. The Council maintains roads and bridges, constructs new bridges and buildings, alters boundaries, decides the number of Parish Councillors, supervises matters of public health, administers the Contagious Diseases (Animals)

Act, appoints a medical officer of health and a sanitary inspector, and may also lend money to parish councils. The Standing Joint-Committee, which controls the police, contains Commissioners of Supply (this was the chief governing body prior to 1889) as well as members of

The Town Hall, Elgin

the County Council, while the sheriff and his substitute are *ex officio* members.

In 1894 a Parish Council was set up in every parish in Scotland. Its most important duty is the administration of the Poor Law, formerly administered by Parochial Boards. The Parish Council has also to levy rates for education, administer the Vaccination Acts, appoint registrars, provide burial grounds, and maintain rights of way. The parishes of the county are:—Alves, Bellie, Birnie, Cromdale, Dallas, Drainie, Duffus, Dyke and Moy, Edinkillie, Elgin, Forres, Kinloss, Knockando, New

Elgin Academy

Spynie, Rafford, Rothes, St Andrews-Lhanbryde, Speymouth, Urquhart.

Since the passing of the Education Act of 1872, each parish has had a School Board. Morayshire has twenty-two school board districts. The school rates, levied by the Parish Council, are supplemented by governmen grants administered by the Scotch Education Depart-ment. No fees are charged, and education is compulsory for all children between five and fourteen years of age. The schools are divided into three classes—primary, intermediate or Higher Grade, and secondary. The intermediate school provides a three years' training. Secondary schools provide a course extending for at least two years beyond the intermediate stage. The successful completion of a secondary curriculum is rounded off by a Leaving Certificate, which admits a pupil to a university. The most important secondary school in the county is Elgin Academy. In arrangements and equipment the academy compares favourably with the best schools in the country. The Victoria School of Science and Art is another important educational institution. There are secondary schools also at Fochabers, Forres, and Grantown; intermediate schools at Hopeman, Lossiemouth, and Rothes.

The county shares a Member of Parliament with Nairnshire. The burgh of Elgin joins with the burghs of Banff, Cullen, Kintore, Inverurie, Macduff, and Peterhead to form a one-member constituency called the Elgin Burghs.

19. Roll of Honour.

Few names of distinction are found in the early history of the county. The only famous writer that can be claimed for Morayshire is Florence Wilson, better known as Florentius Volusenus, who lived four centuries ago. He wrote several philosophical works, but his reputation rests chiefly on a volume written in Latin, *Dialogus de Animi Tranquillitate*. Wilson was a friend of Cardinal Wolsey, Thomas Cromwell, Boece, Buchanan and many other distinguished men of his time.

Lachlan Shaw, minister of Elgin (1744–1774), may be styled the historian of the county. His *History of the Province* (1775) is remarkably full and accurate. William Leslie, fifty years minister of St Andrews-Lhanbryde, published in 1813 a volume on the agriculture of Moray and Nairn, which is of great value not only as an account of the agricultural conditions, but also as an index of the social conditions at the beginning of last century. Though a native of East Lothian, Sir Thomas Dick Lauder was closely connected with Morayshire and lived at Relugas for several years. He was a friend of Sir Walter Scott and modelled his *Wolfe of Badenoch* on the Waverley Novels. His two best books are *Scottish Rivers* and *The Moray Floods*.

The naturalist and sportsman, Charles St John, who died in 1856, was a great admirer of Morayshire. His *Natural History and Sport in Moray* is known to almost every lover of Nature. Dr George Gordon was also a keen naturalist. Born in Urquhart in the year 1801, Gordon became minister of Birnie in 1832, a charge which he held for fifty-seven years. He had an intimate knowledge of the botany, zoology, geology, and archaeology of his native county. He numbered among his friends most of the great scientists of last century—Darwin, Agassiz, Geikie, Huxley, Hooker. Huxley called one of his most wonderful fossil discoveries *Hyperodapedon Gordoni*.

In William Marshall of Keithmore, born at Fochabers in 1748, the county produced a musician whom Burns pronounced "the first composer of Strathspeys of the age." He wrote over 100 strathspeys and a great number of reels. Burns thought so highly of "Miss Admiral

William Marshall

Gordon's Strathspey" that he composed his well-known song "Of a' the airts" to suit the music.

The county has produced many travellers of distinction, the best known being Gordon Cumming of

Altyre. Born in 1820, Cumming at the early age of twenty-three set out from Grahamstown for the hitherto very imperfectly explored region of Bechuanaland, where antelopes, buffaloes, lions, zebras, giraffes and numerous other varieties of animals fell to his gun. While on this expedition he met Dr Livingstone. Later on when the exploits of Gordon Cumming, the "Lion Hunter," were called in question, Livingstone testified to the correctness of his friend's narrative. After five years of South African travel Cumming returned home, bringing with him hundreds of trophies of the chase. An exhibit of some of these was one of the principal sights of the London Exhibition of 1851. By his volume entitled *Five Years of a Hunter's Life in the Far Interior of South Africa*, Cumming immediately and permanently established his fame.

Several eminent soldiers were born in Morayshire. Among these was General Anderson, founder of Anderson's Institution, who as a young man enlisted in the East India Company's service. A still more distinguished soldier was General Sir George Brown, G.C.B., commander of the Light Division in the Crimean War. Brown was born and died at Linkwood, near Elgin. He was present at several of Wellington's victories in the Peninsular War, where he attained renown by his resourcefulness and bravery. General Brown received honours at the hands of Queen Victoria, the Sultan of Turkey, and the Emperor Napoleon. Major Wilson, now immortal as the hero of "Major Wilson's last stand," was a native of Fochabers. The "last stand"

took place in 1893 during the Matabele War. Major
Wilson and a score of British soldiers were surprised by
a large horde of Matabele warriors, but fought bravely
and slew several hundred of the enemy. At length their

Major Wilson of Fochabers

ammunition failing, they were cut down to the last man.
A Matabele warrior thus describes the final scene: "At
last there was but one [Wilson]. He fought on, a grim
smile on his face, and the fire of a devil in his eyes. He

took guns and bullets from his fallen brothers, and fought on to the end. After his cartridges were finished and he could find no more, he stood—silent and alone—waiting for us to come in and make the final thrust." Little wonder that the county is proud to number among her sons this dauntless " Moray loon."

Donald Smith, better known as Lord Strathcona, was born at Forres in 1820. At an early age he entered the service of the Hudson Bay Co., of which he afterwards became Governor. To him more than to any other individual the construction of the Canadian Pacific Railway was due. Lord Strathcona was a munificent benefactor of educational institutions and held the Chancellorship of Aberdeen University and of McGill University. He raised Strathcona's Horse for the South African War. He discharged the onerous duties of High Commissioner of Canada till his death. A man of the highest integrity, a statesman of extraordinary acumen, a fine example of grit, resourcefulness, and energy, Lord Strathcona died in 1914, leaving a record of which Morayshire, Scotland, and indeed the whole British Empire are justly proud.

Mention should also be made of Bishop Andrew de Moravia, founder of Elgin Cathedral ; Gavin Dunbar (1455 ?–1532), Bishop of Aberdeen ; Patrick Grant, Lord Elchies (1690–1754) ; Alexander Adam, LL.D. (1741–1809), Rector of Edinburgh High School and teacher of Sir Walter Scott ; Admiral Sir Robert Calder (1745–1818) ; Sir William Grant (1752–1832), Master of the Rolls ; and H. A. J. Munro (1819–1885), Professor of Latin at Cambridge.

20. THE CHIEF TOWNS AND VILLAGES OF MORAYSHIRE.

(The figures in brackets after each name give the population in 1911, and those at the end of each section are references to pages in the text.)

Alves (pa. 1109) is a small straggling village to the north of Alves Station. A very durable sandstone is quarried in the district for building purposes. Asliesk Castle, two miles south-west of the village, is a ruined baronial fortalice; and near the old military road stood Moray's Cairn, thought to commemorate a battle. Near its site some Lochaber and Danish axes were found. In the neighbourhood is the Knock Hill (335 feet), the traditional meeting-place of Macbeth and the witches. The modern York Tower now stands on its summit. (pp. 10, 27, 29, 43, 79, 83.)

Archiestown is a village near the east end of Knockando parish. At Dellagyle is a cave said to have been used by Macpherson, the notorious freebooter of *Macpherson's Rant*. The Grants of Ramsbottom, the prototypes of the Brothers Cheeryble in Dickens's *Nicholas Nickleby*, were born in Knockando.

Burghead (1595) is a police burgh situated on the Moray Firth. Near it is the village of Cummingston. The sea-board to the breadth of half a mile was once desolated by sand-drift in a similar manner to the Culbin Sands, but it has been reclaimed for pasture or plough. (pp. 6, 23, 27, 35, 50, 51, 53, 54, 63, 79, 81.)

Dallas (pa. 656) is a village in central Morayshire, on the left bank of the Lossie. Loch Dallas and Loch Trevie are near, and many tiny lochs are dotted over the surrounding district. The ruined outworks and moat of Tor Castle, built in 1400, are half a mile from the village. In the churchyard of Dallas stands the old market-cross, a stone-shaft 12 feet high, surmounted by a *fleur-de-lis*. (pp. 21, 83.)

Duffus (pa. 4005) is a village in a very fertile district. The soil is deep and clayey like that of the Carse of Gowrie. From its fertility the parish has been called "The Heart of Morayshire." Duffus Castle, now in ruins, is near the village. Sandstone and limestone are quarried in the vicinity. (pp. 29, 31, 33, 61, 72, 83.)

Dyke (Dyke and Moy, pa. 1020) is a village in the north-west of the county, on the left bank of the Muckle Water. The parish includes the hamlets of Kintessack and Broom of Moy, the latter on the left bank of the Findhorn. (pp. 3, 12, 29, 83.)

Edinkillie (pa. 832) is a hamlet in the west of the county. About three miles to the south is the Knock of Braemoray (1493 feet), from which an extensive view of the surrounding country may be had. The most interesting antiquity in the neighbourhood is the Castle of Dunphail. (pp. 3, 29, 83.)

Elgin (8656), the capital of the county and a royal burgh, is a bright picturesque town on the right bank of the river Lossie. The scenery of the surrounding districts is very fine. In the immediate vicinity are the beautiful Oak Wood and Quarry Wood. Elgin was probably a royal burgh in the time of David I. The royal charter which it received from Alexander II in 1234 is a treasured possession. The transference of the cathedral to Elgin in 1224 was the making of the burgh, and from that time burgh and bishopric worked together harmoniously and prospered until the Reformation made them bitter antagonists. To the patronage of David I and Alexander II in particular, the town owed much of its prosperity. James II visited Elgin in 1457, and James IV in 1490. To-day Elgin is one of the most attractive towns of Scotland. With a fine situation and a salubrious climate, it is largely a residential town. The Cooper Park, presented by Sir George Alexander Cooper, Bart., in 1902, is a favourite resort. In the museum is a valuable and interesting collection of Old Red Sandstone fossils. The cathedral surpasses all the other antiquities in interest. A stone coffin which is within the ruins is said to

Findhorn

have contained the body of King Duncan before the removal of his remains to Iona. To the east of the cathedral there existed, until comparatively recent times, a deep pool, and the place still goes by the name of the Order Pot. This is plainly a corruption for Ordeal Pot, for at this place many an old hag went through the ordeal of water. (pp. 6, 17, 21, 26, 27, 29, 36, 37, 43, 51, 56, 57, 58, 59, 64, 65, 66, 67, 70, 71, 72, 73, 75, 76, 79, 80, 81, 83, 84, 87.)

Findhorn is a small seaport on the peninsula to the east of Findhorn Bay. A branch railway was opened between the village and Kinloss in 1860; but, proving a financial failure, the line was discontinued. (pp. 12, 23, 50, 51.)

Fochabers, a beautiful town on the right bank of the Spey, is an old burgh of barony. Its market-cross, with part ot the "jougs," may still be seen in the grounds of Gordon Castle, where the original village stood till the end of the eighteenth century. Milne's Institution, founded in 1839, by Alexander Milne, a native of the district, is a well-known school. The churchyard of Bellie, two miles north of Fochabers, contains many interesting tombstones. Red deer roamed at will in the vicinity of Fochabers till 1849, when they were ordered to be exterminated by the Duke of Richmond. (pp. 7, 13, 26, 51, 59, 78, 84, 86, 88.)

Forres (4421) vies with Elgin in its pleasant situation and beautiful surroundings. Its dry climate and sheltered position make it a favourite health resort. The town is of great antiquity, and was once the centre of a separate jurisdiction; but, lacking the patronage of the church, it was outstripped by Elgin. It is supposed that the town was made a royal burgh by William the Lyon, but all the old charters have been lost. The oldest charter now extant dates back only to the reign of James IV, and was granted in 1496. The town was burned by the "Wolf of Badenoch." The Witch's Stone marks the place where three witches were put to death. Cluny Hill belongs to the burgh and is used for recreation. On the highest summit is an octagonal tower, 70 feet high, erected in 1806 in honour of Lord Nelson. A magnificent view is to be had from the top. On the southern slope of the hill is a fine hydropathic. The two brothers, Lieutenant-Colonel Alexander Grant and Lieutenant-Colonel Colquhoun Grant, were natives of Forres. The former contributed greatly to the victory of Assaye, while the latter through his conspicuous bravery and ability during the Peninsular War

Forres

became one of the Duke of Wellington's most trusted officers. Here also was born James Dick, founder of the Dick Bequest for the benefit of the parochial schoolmasters of Moray, Banff and Aberdeen. (pp. 6, 7, 21, 29, 31, 32, 33, 36, 43, 51, 56, 59, 61, 78, 79, 83, 84, 89.)

Garmouth, made a burgh of barony in 1587, was an important sea-port up till 1829. Its decay was brought about by the decrease in the timber supplies from the forests in the basin of the Spey, by the use of iron instead of wood for ship-building, and by the vagaries of the Spey. Garmouth was plundered by the Marquis of Montrose in February, 1645, and three months later burned by his orders. (pp. 21, 35, 51, 52, 57.)

Grantown (1451), a fine town 712 feet above sea-level, was founded in 1765 by Sir Ludovick Grant and Mr Grant of Grant, from whom it derives its name. It became a police burgh in 1898. Its elevation, its pine woods, its beautiful scenery of river and moor, account for its popularity as a health and tourist resort. Since the visit of Queen Victoria in 1860, it has yearly become more frequented by holiday-makers. Castle Grant, about two miles to the north, is the ancient residence of the chiefs of the Grants. (pp. 8, 15, 16, 26, 35, 39, 79, 80, 84.)

Hopeman is a village two miles east of Burghead. The harbour is suitable only for small craft. (pp. 35, 50, 51, 79, 84.)

Kingston, a village near the mouth of the Spey, was once a busy port and a thriving shipbuilding centre. Changes in the channel of the Spey, however, made the harbour trouble-some and, later, dangerous. Notwithstanding the cutting of a new channel, the village never regained its former prosperity. (pp. 13, 17, 22, 52, 53.)

Lhanbryde (St Andrews-Lhanbryde, pa. 1183) is a pretty little village on the main road between Elgin and Fochabers. In the north of the parish is Pitgaveny House, which Skene identifies with Bothgowan, where King Duncan was slain. (pp. 21, 29, 79, 83, 85.)

Lossiemouth (4207) is a town situated on the Moray Firth, five miles north of Elgin. The ancient villages of Old Lossie and Seatown lie close to the shore, and at one time contained the whole community. A new harbour was built a little further north. Thither the greater number of the inhabitants migrated and the new village of Branderburgh grew up. The three villages along with Stotfield, which consists almost entirely of villas, constitute the burgh of Lossiemouth and Branderburgh. Lossiemouth has recently attained great popularity as a watering-place. (pp. 7, 17, 21, 22, 23, 27, 35, 42, 43, 50, 51, 79, 81, 84.)

Rafford is a small village to the south-east of Forres. There is a tradition that in 954 Malcolm I was slain in the vicinity, at Blervie. (pp. 21, 29, 75, 83.)

Rothes (pa. 1350) is a police burgh on the left bank of the Spey. In the neighbourhood are two fine waterfalls, the Doonie Linn and the Drumbain Linn. Distilling is the principal industry, the distilleries of Glengrant, Glen Rothes and Glen Spey being near the town. Other industries are saw-milling and copper-working. The haugh land is well cultivated and the surrounding heights, including the Conerock, are beautifully wooded. Doubtless the haugh of Rothes was laid down at the bottom of a large lake in the course of the Spey before the river succeeded in cutting through the rock at Sourden. Rothes suffered great devastation in the floods of 1829. The haugh land was greatly damaged by the deposition of sand. In one house in the village a salmon weighing six pounds was caught. On the summit of a round hill adjoining the town are the ruins of Rothes Castle. The castle is of great antiquity and was surrounded by a fosse. In it Edward I took up his quarters on July 29, 1296, as he returned from Elgin. (pp. 3, 7, 16, 26, 35, 43, 79, 80, 81, 83, 84.)

NAIRNSHIRE

1. County and Shire[1]. Origin and Administration of Nairnshire.

For many centuries Nairnshire formed part of the extensive Province of Moray, which in the eleventh century stretched along the sea-shore from the mouth of the Spey to the Dornoch Firth with the hinterland to the source of the Spey and to the watershed from about Glen Quoich to Loch Broom. The county of Nairn proper may be regarded as dating from near the end of the twelfth century, when William the Lyon raised it into a separate sheriffdom. From time to time the boundaries have undergone considerable alterations. In 1891 the Boundary Commissioners cut off several detached portions for County Council purposes. Thus in the civil parishes of Cawdor and Croy and Dalcross in 1911 there was a population of 288 under Nairnshire for parliamentary purposes but under Inverness-shire for County Council administration.

The county takes its name from the town of Nairn. The old name of Nairn was Invernarne, i.e. the mouth of the Nairn. The word Nairn is derived from the Gaelic *uisge na-fhearna*, "the water of alders." As numerous alder trees grew, and still grow, along the banks of the river, the name is highly appropriate.

[1] See p. 1.

The first heritable Sheriff of Nairnshire whose name has come down to us was Andrew, Thane of Cawdor, who died in 1405. After 1747 the office of Sheriff was no longer hereditary. Nairnshire is under the same Sheriff as the counties of Inverness and Moray.

In 1782 a Lord-Lieutenant was appointed for each county. The Lord-Lieutenant of Nairnshire is assisted by thirteen Deputy-Lieutenants and by Justices of the Peace.

Like the other Scottish counties, Nairnshire has its County Council, while each parish has its Parish Council[1]. The four parishes entirely within the county are Nairn, Auldearn, Cawdor, and Ardclach. Croy and Dalcross is partly in Nairnshire, partly in Inverness-shire. Each parish has a School Board, and there are six school board districts in the county, Nairn having both a burgh and a landward district. Rose's Academical Institution or, as it is generally called, Nairn Academy, the only secondary school in the county, has a record of which Nairnshire may well be proud.

The county unites with Morayshire in sending one member to parliament. The burgh of Nairn is represented by the member for the Inverness burghs.

2. General Characteristics. Climate. Communications.

Scotland is usually said to consist of three distinct natural divisions—the Northern Highlands, the Central Lowlands, and the Southern Uplands. But the area of

[1] See p. 81.

which Nairnshire forms a part—the lowlands on the
Moray Firth—does not fall very well within any of these
divisions. For meteorological, topographical, and geological
reasons we may look upon this region as a separate geographical unit which, for want of a better designation, we
may call the Northern Lowlands.

For agricultural purposes the county is particularly
well-favoured. The climate, especially of the seaboard
plain, is dry and temperate, while much of the soil is
extremely rich. As part of the Province of Moray,
Nairnshire shares in the many eulogiums passed by early
travellers on that "pleasant and plentiful country."

The mildness of the climate is especially noteworthy
when we consider that the county lies between the
57th and 58th parallels of north latitude. The mean
temperature and rainfall of the town of Nairn for each
month of the year are as follows :

	Mean Temperature. Degrees Fahrenheit.	Mean Rainfall. Inches.
Jan.	37·3	2·0
Feb.	37·9	1·6
Mar.	39·7	1·7
Apr.	44·2	1·3
May	49·1	1·7
June	54·9	1·8
July	57·4	2·7
Aug.	56·9	2·7
Sept.	52·9	2·5
Oct.	46·0	2·4
Nov.	41·0	2·1
Dec.	37·6	2·2

The low rainfall, the infrequency of snow, and the uniform geniality of the weather will be found explained on page 39.

The county is bounded on the north by the Moray Firth, on the east by Morayshire, and on the south and west by Inverness-shire. It consists of two well-defined natural divisions. In the north is a seaboard plain which runs inland (gradually increasing in height above sea-level)

Nairn from the Railway Bridge

for about four miles. In this area, which may be regarded as the westward continuation of the Laigh of Moray, farming is carried on with much success. Then there is a series of comparatively low hills, the two chief eminences being the Lethen Bar and the Hill of Urchany. South of this is a stretch of highland country abounding in glens and straths, and flanked by lofty hills. Here crofting and stock-raising are the chief employments of the people.

The rivers flow in a north-easterly direction. Occasionally great damage is done by flooding, the most notable instance being that of the "Moray Floods" in the year 1829[1].

By far the greater part of the population is found in the northern plain. Indeed, the burgh of Nairn alone accounts for rather more than half the population of the county.

The main lines of communication for Nairnshire are the same as those of Morayshire, described on pp. 78–81.

Up till 1805, when the stage-coach was first employed, the riding post to Aberdeen was the only means of communication between Nairn and the south. The first railway in the county was that connecting Nairn with Inverness, and was opened in 1855. Three years later it was connected with the Aberdeen Junction Railway and eventually it was merged in the Highland Railway. The advent of the railway brought prosperity to the town of Nairn. Every year larger numbers of tourists visited "the Brighton of the North."

3. Size. Shape. Boundaries.

In size Nairnshire ranks thirtieth among the thirty-three counties of Scotland. Its total area is 104,252 acres. Inverness-shire, the largest county, is more than twenty-five times the area of Nairn. The length of the

[1] See p. 59.

county from the mouth of the Nairn to Carn Glas in the extreme south is $17\frac{1}{2}$ miles and its average breadth is about 11 miles. The coast line measures rather more than 9 miles.

In shape the county may be regarded as a pentagon, two sides of which go to form the eastern boundary.

The boundaries are largely artificial. Starting at the north-east corner at the middle of The Bar, the boundary follows an irregular course south by east to the Muckle Burn, near Earlsmill. Running along that stream for about two miles, it again strikes in a general south-by-east direction till it reaches the Findhorn. For about a mile the river forms the boundary, which then runs south-east to the high ground, between the Dorbock and the Findhorn and their tributaries. It follows the high ground (with one or two minor deviations) in a general south-by-west direction to Carn Glas. From this summit the line turns north-westward to the southern end of the Streens, where it again touches the Findhorn. Going up the river for about three-quarters of a mile, it then runs in a north-westerly direction, following the watershed between the tributaries of the Nairn and the Moy Burn, and passing Carn nan tri-tighearnan (2013 feet), till it touches the river Nairn at the most westerly point in the county in the neighbourhood of Croygorston. For the next three miles the boundary runs along the Nairn, and thereafter its course northward is far too irregular and arbitrary to be minutely described here. It reaches the Firth about four miles east of Fort George. The Moray Firth forms the entire northern boundary.

4. Surface and General Features.

The configuration of the county falls into two divisions. Adjoining the Moray Firth is an extensive plain which constitutes the lowland district. In striking contrast to this low-lying or gently undulating ground, the southern and by far the larger part is a very hilly region.

Much of the seaboard is lined with sand-dunes. In the north-east are the Maviston Sand Hills, which afford a magnificent view of land and sea. The sand which overwhelmed this area had probably been cast up on the shore of the county by the westward tidal current, and was afterwards driven north-eastwards by the prevailing south-west winds. A short distance south-east of Lochloy is the site of the extinct fishing village of Maviston, the name by which the locality still goes. The fishermen of Maviston may have used the now silted-up "port of Lochloy" for their boats while their homes and crofts were at Maviston. About a mile west of Nairn may be seen a fine example of a kame of shingle and sand, which extends to the neighbourhood of Fort George, with an average height above sea-level of 120 feet.

To the south of the sandy seaboard lies a plain of considerable size. This is a fertile and picturesque strip of land, stretching inland for about four miles, and drained by the river Nairn and the Muckle Burn. In the middle of the county a height of 600 feet is reached. Most of the land is cultivated or under thriving plantations. The waste land consists largely of peat hags, and the drier parts are bright with clumps of whin. The region is drained by the Muckle Burn.

To the south lies a typical highland district. The principal hills are two spurs from the Grampians, cleft and drained by the Findhorn. It is a region of heather-clad moorland surrounded by lofty hills, the highest of which is Carn Glas (2162 feet). Some of the other heights are: Carn Sgumain (1370), Maol an Tailleir (1373), Hill of Aitnoch (1351), Carn a Gharbh Ghlaic (1523), Carn nan tri-tighearnan (2013), Carn na Sguabaich (1522), Carn Allt Laoigh (1872). Grouse moors are numerous and a large part of the area is devoted to grazing. The scantiness of the population is accounted for by the sterility of the soil and the moist climate.

5. Watershed. Rivers. Lakes. Coast-line.

The general slope of the county is to the north and east, so that the rivers flow in a north-easterly direction. The two spurs from the Grampians which enter the county form the principal watersheds. The northern slope of the more southerly spur drains into the Findhorn by several streams, including the Rhilean Burn, the Leonach Burn, and the Tomlachlan Burn. The southern side of the northerly spur, including Carn nan tri-tigh-earnan, is also drained by the Findhorn; while the rainfall of the northern side and of practically all the county to the north of it finds its way to the Firth by

Glenferness

the river Nairn and the Muckle Burn, with their numerous tributaries.

The Findhorn rises in the Monagh Lea Hills in Inverness-shire and has a course of nineteen miles through Nairnshire. Entering the county by a narrow defile at Pollochaig in the south-west, the river flows in a north-easterly direction. The first part of its course in the county lies through the Streens, a highly picturesque district. At Dulsie the river has hewn out a yawning chasm, which is spanned by a bridge of one arch constructed by General Wade in connection with his military road from Grantown to Fort George (see p. 113). The surrounding scenery is wild in the extreme, and for the next ten miles the course of the river is very romantic. At Ferness, about two miles below Dulsie Bridge, the Findhorn sweeps round a rocky peninsula, and here the scenery is of the grandest description. Many of the rocks rise in fantastic shapes to a great height sheer from the river-bed, and the surrounding country is finely wooded with pine and birch. After leaving Ferness the river passes through a beautiful tract of country and dashes into Morayshire near Downduff.

The river Nairn rises in Inverness-shire at a height of 2600 feet above sea-level. Of its total course of 38 miles, 12 miles are in Nairnshire. It flows in a north-easterly direction and empties its waters into the Moray Firth at the county town. Touching Nairnshire for the first time in the vicinity of Culloden Moor, the river for three miles forms the boundary between Inverness and Nairn, and then becomes wholly a Nairnshire stream. It passes through and beautifies the policies of Holme and Kilravock. Then its course lies through the lowlands, where may be seen numerous well-cultivated farms and large tracts of thriving timber. On the right bank the Nairn receives the Cawdor Burn, its only considerable tributary in the county. Some years ago gold was discovered in the Nairn, in the sand near the place where the Daltullich Burn joins the river. Unfortunately, the experts who

The River Nairn

investigated the matter came to the conclusion that the gold was not present in sufficient quantity to repay the cost of search. The fishing on the Nairn includes trout and salmon.

The river Nairn has made several attempts to alter its course so as not to touch the town of Nairn at all. In 1734 the Council finding that the river entered the sea to the *west* of the town agreed to take measures to "rectify the same" with all possible speed. In 1777 the river discharged into the Firth about a mile *east* of the town, but since 1820 it has stuck faithfully to its present bed.

There are few lochs in the county, and all of them are small in size. In the north-east are situated Cran Loch and Lochloy. The latter is separated from the sea by a marsh, although formerly it was an arm of the Moray Firth, for as early as 1196 we find references to the "port of Lochloy." At the west end of the loch stood the village of Lochloy, of which now no trace remains. Querns and hearthstones have been found; and the site of an old church is a little farther west.

Almost four miles south-west of the town of Nairn and to the south of the great kame, lies the Loch of the Clans (see p. 125). Formerly of considerable extent, it is now a tiny sheet of water owing to drainage operations. The chief interest of the loch is the fine example it furnishes of a lake dwelling. Loch Flemington, a mile and a half south-west of the Loch of the Clans, is cut in two by the boundary line between Nairn and Inverness. Loch Belivat lies about eight miles from Nairn on the Grantown road.

The coast-line of the county is very regular, the prevailing rock being sandstone. Close to the boundary between Nairn and Moray are the extensive Maviston Sandhills. Opposite the sandhills is The Bar, the western half of which belongs to Nairnshire. The Bar marks the place where the Findhorn formerly discharged into the

Firth, and in summer is frequented by vast colonies of birds, including a few rare species. Nairn links, situated to the west of the town, are well known to lovers of golf. Not only is the course one of the finest and also one of the oldest in the country, but the view obtained from it is superb. The Ord of Caithness, Dunrobin, the Souters of Cromarty, and Ben Wyvis are all clearly visible. In the extreme west the coast-line consists of a long narrow peninsula which points westward. It has been formed from material laid down by the tidal current from the east. A small portion of the peninsula belongs to Inverness-shire.

6. Geology[1] and Soil.

Geologically, the county falls into two well-marked portions. The low-lying plain in the north is composed of the Old Red Sandstone, which at one time covered practically the whole shire, while the southern and larger part of the county is characterized by metamorphic and crystalline rocks.

Along the shore in the northern plain the sandstone is quarried at Kingsteps and other places. The fossils found belong to the period of the Upper Old Red Sandstone, and this division lies unconformably on the middle series. In the upper division shales, and clays, and obliquely-bedded sandstones are found. About a century ago a fossil-fish bed was discovered at Lethen Bar in the parish of Ardclach. The fossils are of great scientific

[1] See p. 23.

interest and value. It is interesting to note that the plates for Agassiz' great work on the fauna of the Old Red Sandstone were taken from Lethen Bar fossils. In addition to the fine kame which rests on a plain of sand between Flemington, near Fort George, and Meikle Kildrummie, smaller kames are found on the moors in the uplands of the county. Many erratic blocks occur in the middle and the northern parts. One of these, called "Tomriach," is found at Croygorston and is supposed to be the second largest block of the kind in Scotland. It is 13 feet high, 35 feet long, 20 feet broad, and fully 90 feet in circumference. Its weight is estimated at 550 tons. The parent rock, from which this boulder and most of the other travelled boulders in the county are supposed to have come, lies thirty miles to the south-west. The position of the boulders seems to indicate that the ice-sheet moved in an easterly direction along the hill-slopes. At Newton also the ice-markings show that the ice-sheet moved in this direction. Glacial deposits are so widely spread that few rock surfaces have been left exposed. These deposits comprise a lower boulder clay and an upper boulder clay separated by an important series of gravels and clays. Near the coast traces of marine terraces occur at heights of 25, 50, and 100 feet, and even higher, above the present sea-level.

The south of the county is composed mainly of what is known as younger highland schists, with which granite masses are associated. All the higher ground consists of these schists or of granite. The schistose rocks comprise micaceous gneisses, mica-schists, and quartzites. A mass

of porphyritic gneiss is found on Carn nan tri-tighearnan.
Numerous dykes of granite intersect the schists. The
largest granite mass extends from the neighbourhood of
Lethen Bar, by way of Glenferness to the Bridge of
Dulsie and beyond. The second mass is found on the
slopes of Beinn Bhuidhe Mhor and Beinn nan Creagan,
in the extreme west of the county. A less extensive

The Bridge of Dulsie

mass consisting of a beautiful porphyritic granite occurs
near Rait Castle, and is quarried for building purposes
and ornamental work at Kinsteary, not very far from the
town of Nairn.

In the northern part of the county the soil is generally
loam with the exception of a narrow sandy strip along the
shore. Some parts, formerly moorland but now cultivated,

are a sandy gravel. The mossy soil is almost entirely restricted to the parish of Ardclach. In the highland part of the county the plains along the streams are covered with a sandy soil or a loam with a considerable admixture of sand.

7. Natural History[1].

The fauna of the county does not differ greatly from that of Scotland as a whole. Of the carnivores, stoats and weasels are fairly common. The fox and the otter are occasionally met with. A family of badgers was seen in the county as recently as 1891, and in the same year a polecat was discovered at Ferness. The rats and other rodents of Nairnshire, the moles and other insectivores, and the bats are generally identical with those of Morayshire.

Fully 200 species of birds are recorded for Nairnshire. The common buzzard, the rough-legged buzzard, the hen-harrier, and the peregrine falcon are now rarely seen. The Polish swan, the common crane, the shrike, the sand grouse, and the gray phalarope have been noted in the county. The cross-bill frequents the valley of the Nairn. Several kinds of owls are known, two of which—the tawny owl and the long-eared owl—are resident. The kingfisher, the redstart, the fieldfare, the redwing, and the golden-crested wren are all found. Among the game birds of the county are grouse, partridges and capercailzies. Wild duck frequent the marshes and snipe are numerous. The kestrel,

[1] See p. 29.

the sparrow-hawk and the merlin occur. Black-headed gulls are abundant. The wood-pigeon, the rook, the sparrow, and the starling are regarded as pests.

The reptiles and amphibians of the county include viviparous lizards, palmated newts, toads and frogs.

The flora is practically the same as that of Morayshire, described on pp. 33–35. Many varieties of ferns, including the adder's tongue and the moonwort, are seen. Among the rare plants of the county may be mentioned the twayblade, the bird's-nest orchis, the bird's nest, and the lily of the valley, all of which flourish in the Cawdor Woods. The yellow gagea grew at one time on the Blackhills near Auldearn. The chaffweed can be seen at Maviston; the water pimpernel and the water lobelia by the sides of Lochloy. The county contains many fine woods. The principal varieties of trees are Scots fir, larch, and spruce. Extensive plantations are found on the estates of Brodie, Cawdor, Kilravock, and Lethen.

8. The People — Race. Language. Population.

What has been already said (p. 40) about the race and the language of Morayshire, applies equally to the inhabitants of Nairnshire. The one exception is in the place-names. All over the shire Celtic place-names predominate. In 1911, 929 inhabitants, or 9·9 per cent. of the entire population, as against 1335 in 1901, were returned as able to speak Gaelic. Of the 929 Gaelic speakers only 156 were born in the county, while 410 were born in Inverness-shire, and 263 in Ross and Cromarty.

Nairnshire has a sparse population, the census of 1911 giving a return of 9319 inhabitants. In 1801 the census return was 8322, compared with which the present popu-

The High Street, Nairn

lation shows an increase of 1062 or 12·9 per cent. But to-day the county, owing to changes in its boundaries, is smaller than in 1801. Making allowance for these alterations, we find that the present county of Nairn shows an increase of 1920 or 23·3 per cent. over the population of the corresponding area in 1801. The maximum population was in 1881, when it was 278 or 2·7 per cent. higher than the present (adjusted) population. The density in 1911 was 57 to the square mile. Lanarkshire, the most densely peopled county in Scotland, has 1633 to the square mile, while Sutherlandshire, our most thinly peopled county, has 10 to the square mile. Fully half of the population of Nairnshire is found in the burgh of Nairn : 4661 within the burgh, 4658 outside. Compared with 1901, the burgh shows an increase of 174 or 3·9 per cent., the extra-burghal population a decrease of 146 or 3·0 per cent. The increase in Nairn may safely be attributed to its popularity as a seaside and health resort and to its excellent golf-links.

The population of the landward part of the parish of Nairn is 1245, exactly the same as in 1901. The population of Auldearn shows a decrease of 44, Cawdor a decrease of 78, and Ardclach a decrease of 13, as compared with 1901. It will thus be seen that there is a distinct decrease in the rural population, largely due to migration to manufacturing centres, to emigration, to the cheapness of imported food supplies, and to the introduction of labour-saving agricultural machinery.

Agriculture gives employment to 988 males or 36·1 per cent. of the occupied males, and of these 227 are farmers or crofters. Fishermen number 298; while 175 males are connected with the building trades. Of the female population 626 are employed in domestic service, 97 in agricultural work ; 131 are fishcurers, 62 are dressmakers, and 51 are teachers.

9. Agriculture[1] and other Industries.

Agriculture is the only really important industry, and in the northern part of the county where suitable conditions prevail it has been brought to great perfection. In the uplands of the county different conditions prevail. The land is poorer and the climate is cold and damp. In the parish of Ardclach most of the soil is of a mossy cast. The inhabitants of this less-favoured area devote their attention largely to pastoral pursuits.

The improvement of the agriculture of the county is due in great measure to the Nairnshire Farming Society, founded in 1798, which had for its object the promotion of "improvements in agriculture both by precept and example." The society was energetic and practical to a degree. It discussed the all-important question of the winter feeding of cattle; the members reported any successful experiments for the improvement of agriculture; prizes were given to encourage the growing of turnips and potatoes; experiments were made in liming the land; ploughing matches and cattle shows were held; corn markets and feeing markets were established.

The county contains 104,252 acres of land, 25,344 of which were under crops and grass in 1912. That is, only 25 per cent. of the total acreage is under cultivation, which is exactly the average percentage of cultivated land throughout Scotland. The uncultivated part consists of

[1] See p. 44.

hills, moors and woodlands. 13,335 acres, or 12 per cent. of the total area, are under wood, the plantations consisting chiefly of Scots fir, larch and spruce.

As a rule the farms are small.

In 1912 the acreage under the principal crops was as follows :

Oats	...	...	...	5749	acres
Barley	...	...	...	2789	,,
Turnips	...	...	...	3988	,,
Clover, Grasses, etc.		...		9405	,,
Potatoes	...	...	...	328	,,
Rye	...	...	...	86	,,
Wheat	...	...	...	76	,,

The large quantities of oats, turnips, and clover are required for the live-stock industry, while the distilling industry of the county, and still more of Morayshire, accounts for the relatively large amount of barley. Very little wheat is now grown in the county. This is specially remarkable, since up till the middle of last century this crop was raised on practically all the best soil. At the International Exhibition in London in 1851 the first prize was awarded to a sample of wheat grown in Cawdor. But the return of wheat per acre was far too small to prove remunerative when the price fell, and consequently the farmers had to turn their attention to other crops.

In the improvement of stock the Nairnshire farmers have not lagged behind their enterprising Morayshire neighbours. Excellent herds of cattle are found in the northern part of the county. In 1912 the number of cattle was 6025 ; of sheep 16,131 ; of horses 1371 ; and of pigs 685. Very fine specimens of Clydesdale horses may be seen at the annual shows.

Fishing is prosecuted with considerable success from the county town, where it is the most important industry. Haddocks and flounders are the principal kinds of fish. Drift nets, set nets, and lines are used. Nearly 100 boats are engaged, some of them being as fine as any on the Moray Firth. According to the season of the year

Nairn Harbour

Nairn boats are to be found along the east coast from Lerwick to Yarmouth. From December to April the fleet is at home, busily engaged in the white fishing. The "fishers" live in a part of the town by themselves and mix very little with the other inhabitants. They are intelligent, hardworking, and thrifty. The surname Main is extremely common among them.

Brackla Distillery was erected fully a century ago and has been worked with ability and success. It proved a great boon to the farmers, furnishing a local market for their barley, and thus saving them the inconvenience of transporting it to Morayshire or Inverness-shire. Recently new machinery has been introduced, the power now being got both from water and from steam. Peat is largely employed for heating, and all the barley used is home-grown. A pure malt whisky is produced.

The county possesses little mineral wealth. In 1909 the only minerals raised were 200 tons of brick clay, 2558 tons of igneous rocks and 3520 tons of sandstone. The principal sandstone quarries are found along the shore in the neighbourhood of Kingsteps.

10. History[1].

The history of Nairnshire in the early centuries of our era is obscure, but the district certainly formed part of the dominion of the Northern Picts. The introduction of Christianity by Columba marks the beginning of the authentic history of the county. At Auldearn, the name of Columba survived until the year 1880 in St Colm's fair, which was held annually on 20th June. At Clava and at Urchany chapels belonging to the early Columban Church were built close to existing cairns and circles. Celtic church monuments are also found in the county. The Celtic Church remained the most powerful force in the district for 150 years, and during that period it not

[1] See p. 53.

only propagated religious doctrines but also served as a centre for the spread of education, art, and industry.

For the next five centuries the story of Nairnshire and the story of Moray are inseparably linked together. What is true of the one is true of the other.

Edward I's visits to Elgin were not unfelt in Nairn-shire. On his first, in 1296, he issued summonses for the Parliament of Berwick, at which Nairnshire was repre-sented by Sir Gervaise de Raite, Sir Andrew de Raite, and Sir Reginald Chien. Edward was again in Moray-land in 1303, and stayed a month in the Castle of Lochindorb. It was no easy task to find rations for his vast army in that unproductive region. Supplies were brought from as far north as Dingwall, while the sheriff of Nairn forwarded 26 cattle, 26 sheep, and 40 pigs as food for the garrison.

The men of Nairn had fought with Sir Andrew Moray in support of Wallace ; they now shared in the struggle that ended at Bannockburn, and in the honours and privileges of the new earldom of Moray bestowed by Bruce on his nephew Thomas Randolph.

From this time nothing of striking importance can be connected specially with Nairnshire till the seventeenth century, when, during the Covenanting "troubles," it was the scene of one of Montrose's brilliant victories. Learning that the Royalist army was encamped at Auldearn, Hurry, at the head of 3500 foot and 400 horse, left Inverness with the object of surprising and crushing Montrose, whom he attacked, May 9, 1645. For long the battle was doubtful, but when Macdonell, contrary to Montrose's express commands, left his defences it looked as if the Covenanters were to win. Macdonell was reported to be completely routed. With characteristic resourcefulness, Montrose exclaimed, "What! Macdonell gaining the victory single-handed! Come, my Lord Gordon, is he to be allowed to carry all before him and leave no laurels for the house of Huntly?" The gallant youth at once led his small band of horsemen to Macdonell's assistance and converted disaster into victory. Montrose is said to have lost 200 men, the Covenanters 800.

A century later Nairnshire was the scene of one of the final events in the Jacobite Rebellion. The Duke of Cumberland crossed the Spey on April 12, 1746; and the Duke of Perth and Lord John Drummond giving way, he reached Nairn two days later. Prince Charles Edward drew up his hungry and discontented men on Drummossie or Culloden Moor, April 15. That was Cumberland's birthday and the Jacobite leaders anticipated that merry-making would render the royalist watch less vigilant. Accordingly a resolution was come to, which if successful might have changed the destinies of our country. In spite of the opposition of the chiefs, who laid stress on the physical weakness of their men, it was resolved to attempt a night surprise on Cumberland's camp. At eight in the evening the men were roused and started on their twelve-mile march. But exhausted through want of food, bewildered in the dark, delayed by the rough road, they found themselves still three miles from Cumberland's camp when day began to break. The royalist troops, also, were on the alert. It was clear that the attempt had failed; and the footsore Highlanders toiled back with heavy hearts to await their fate on Drummossie Moor.

In 1829 Nairnshire, in common with Morayshire, suffered a devastation without parallel in the history of Scotland. In that year the "Moray Floods" took place. The Nairn was greatly swollen. Dwelling-houses were demolished; arable land and forest were carried off to the sea by the acre; crofters watched their homesteads as they sailed towards the Firth. The damage due to the floods was enormous. Lord Cawdor's loss, for example, amounted to fully £8000.

11. Antiquities.

The county is extremely rich in prehistoric remains. Relics of settlers who used stone implements are found all over the district. At Slagachorrie, some four miles south of Nairn, numerous stone axes and flints have been discovered, some of them remarkably fine. One flint is beautifully serrated, having no fewer than thirty teeth in a space of one inch. These are the earliest examples of craftsmanship that have come down from prehistoric times.

Remains of cairns and stone circles abound in the district. Examples of these are, or were formerly, found at various places including Auldearn, Urchany, and Moyness. In the circles urns have been discovered, thus showing that the people who built the circles, if they also buried the urns, knew something of the potter's art

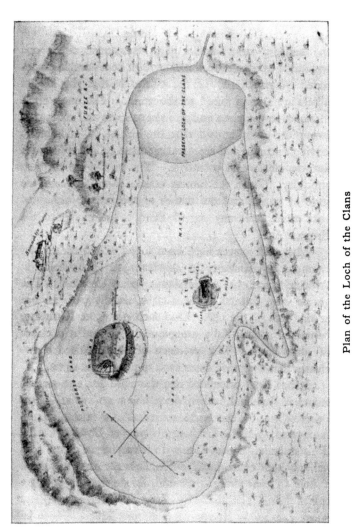

Plan of the Loch of the Clans

and practised cremation. The circles further show that
the builders were familiar with concentric circles and
were able to take measurements with considerable exactitude. A peculiarity of the circles is that the highest
pillars are always found to the south. The outer circle
of stones shows noon and also the solstices and equinoxes.
From this it is argued that the circles, in addition to
serving as tombs and representing some religious idea, did
duty as clocks and calendars. The circles are of great
antiquity. From the bronze articles found in them, the
style of burial, and the nature of the pottery associated
with them, we may regard them as belonging to the
Bronze Age.

Many cist graves have been found in the county, for
example, at Kinsteary in Auldearn, at Mid-Fleenas in
Ardclach, and at Auchindoune near Cawdor. The
graves were formed of three slabs, usually with an urn
at one end of the bed. One such grave, at Kinsteary,
contained a beautifully ornamented vase and a necklace
of glass beads. The presence of the necklace shows
clearly that the grave dates back to prehistoric times.

It has been calculated that there are 204 cup-marked
stones in Scotland, and of those about 40 are in Nairnshire. One of the most remarkable of these stones,
discovered at Little Urchany, is now at Cawdor Castle.
This stone has a dozen cups on it varying in depth from
half an inch to an inch, and in diameter from three and
a half to two inches. Several fine examples of these
stones have been found at Barevan Churchyard; and
another, a highly interesting one from Moyness, is to be

seen at Cawdor Castle.   No satisfactory theory as to the use of the stones has yet been propounded.

In 1863 a fine example of a lake-dwelling or crannog was discovered at the Loch of the Clans.   An artificial island had been made in the loch by forming a raft of logs, upon which stones and other logs were deposited. The lower beams were so constructed as to keep out the water.   A turf-roof would complete the structure. In a marsh about fifty yards distant was the foundation of another lake-dwelling.

Vitrified forts, where the stones have been run together by the application of heat, are found at Dunevan, Dunearn, and Castle Finlay.   Dunevan, a hill 800 feet high in the parish of Cawdor, has from its summit a wide view, extending on the north to the Moray Firth. Dunearn fort is situated on the edge of an extensive moor on the right bank of the Findhorn.   The summit of the fort, extending to two acres, has been brought under cultivation and there is very little evidence of vitrification now to be seen.   Like Dunevan, the fort would serve as an excellent retreat for the inhabitants of the valley in time of danger.   Castle Finlay differs from the others in being situated at the foot of a hill.   Its situation shows that such forts were not merely sites of beacon fires but also strongholds.

12.   Architecture.

Fortifications built of stone and lime were probably first introduced into Scotland by the Normans in the eleventh or the twelfth century. They were quadrangular buildings, called keeps, and most of the Scottish strongholds were modelled on them. Kilravock Castle is an example of a keep enlarged by later additions. The keep was built in the fifteenth century. Two centuries later it was enlarged into a castle ; and still later equipped to suit modern ideas.

Rait Castle

Rait Castle, situated at the foot of the Hill of the Ord, about three miles south of Nairn, has proved a great puzzle to antiquarians and architects, owing to the very exceptional modifications of the keep plan. Thus, the windows are pointed and show Gothic arches and mullions. In many ways the building suggests ecclesiastical rather than castellated architecture. On a close examination, however, distinct traces of the grooves of the portcullis are seen. Further evidence of the defensive equipment is found in the fact that the tower was so situated as to protect from the south, the side most open to assault. The castle is believed to have been erected early in the fifteenth century.

The keep plan gave comparatively little accommodation and was badly suited for defence against firearms. Accordingly, Z plans, with two towers at diagonally opposite corners, were introduced. Inshoch Castle, dating from the sixteenth century, but now a ruin, about half-way between Nairn and Brodie railway stations, is built on this plan. The towers are round externally. The south-west turret, the larger one, contains a square apartment with stone seats, while the south-east one is so small that it contains nothing but the wheel stair leading to the first floor. On this floor is the hall, 30 feet long and 17 feet broad.

Cawdor Castle, the ancient seat of the Thanes of Cawdor, is a fine example of a mediaeval fortress. Permission to build the castle was granted to the Calders of Calder in 1454, but most probably a still more ancient keep existed on the site, for the Exchequer Rolls for 1398 refer to expenses connected with "Calder Castle." Though parts are thus of great antiquity, most of the present building belongs to the seventeenth century. The numerous additions and alterations have been carried out so well that the ancient character is faithfully maintained, and the building has been well described as "a relic of the work of several ages." The massive central keep, three storeys high, is 45 feet long and 34 feet wide, while its walls carry battlements and towers far higher than the more recent buildings. There is a deep moat, with the drawbridge still remaining. The gateway, which leads to a small courtyard, is protected by a massive iron gate

Cawdor Castle

and loop-holed walls, and is crowned with a belfry. The castle stands on the edge of a rock on the right bank of the Cawdor Burn. A wall of enceinte was built on the south side and surrounded the original keep, but some portions of the wall were utilised in later additions to the fortalice.

An interesting ecclesiastical ruin is that of St Ewen at Barevan. The walls, which are still intact, enclose a quadrangular space 70 feet in length and 20 in breadth. The style of architecture is that of the First Pointed period. The chapel was used up till 1619.

Among the modern public buildings of the town of Nairn may be mentioned the Public Hall, which includes the interesting Nairn Museum and the Literary Institute and Reading Room; Nairn Academy; and the Town and County Hospital.

13. Roll of Honour.

The roll of honour of the county contains no names famous in literature, science, or art. The genius of the inhabitants seems to have run in other directions. It is seen at its best in deeds of derring-do and so the most famous men of the county have been soldiers or travellers.

Among the distinguished soldiers of the county Lieutenant-General Gordon occupies a prominent place. He saw active service at Malta, at Gibraltar, at Walcheren, in the Peninsula and in France. His conspicuous bravery at the crossing of the Nive, where he was wounded and had his horse shot under him, led to his appointment as Colonel of the 54th Regiment. After the fall of Napoleon Gordon settled down in Nairn, where he lived to see five of his sons make the army their profession. He died in 1856. About the time of General Gordon's residence

at Nairn a large number of Nairnshire lads joined the
army, and many of them attained distinction. One of
these was Patrick Grant, who, after a brilliant military
career, was made a Field Marshal. A similar honour and

Colonel J. A. Grant

also a Peerage fell to Hugh Rose. Another brilliant
military career was that of Herbert Macpherson. At the
Relief of Lucknow he won the Victoria Cross; and he
was in command of a Brigade under General Roberts in

the Afghan War. He shared in the memorable march from Kabul to Kandahar; and in the final struggle he and his mixed force of Highlanders and Ghoorkas decided the issue. For his distinguished services Macpherson was made a K.C.B. Next he saw service in Egypt, taking part in the battle of Tel-el-Kebir, and occupying Cairo—an exploit for which he afterwards received the thanks of both Houses of Parliament.

Probably the best known son of Nairn is James Augustus Grant, born in 1829. Grant saw a considerable amount of active service in India and was in Lucknow for two months during the siege. Returning to England on sick-leave, he met Captain Speke who had just returned from Africa after the discovery of Victoria Nyanza. Grant volunteered to accompany Speke, who was returning to Africa to settle conclusively the source of the Nile. The expedition, consisting of over 300 souls, set out for the interior from the African coast in October, 1860. The journey was perilous in the extreme. Sickness was rampant; the climate was unhealthy; the natives were treacherous; food was often scarce; and resources were scanty. Still the expedition was a great success. The White Nile was seen dashing over the Ripon Falls; thus Speke's surmise proved correct. In June, 1863, Grant returned to England and found himself famous. Next year he published an account of his travels entitled *A Walk Across Africa*.

14. TOWNS AND VILLAGES OF NAIRN-SHIRE.

(The figures in brackets after each name give the population in 1911, and those at the end of each section are references to the pages in the text.)

Auldearn (pa. 1248) is a burgh of barony three miles east-south-east of the county town. The rocks of the district are Old Red Sandstone and have been extensively quarried. (pp. 100, 115, 117, 121, 122, 124, 126.)

Cawdor (pa. 847) is a village on Cawdor Burn. Near it are the remains of a vitrified fort and of Barevan's Church, and, most interesting of all, Cawdor Castle. In the castle is a bedroom which is one of the places tradition identifies as the actual scene of King Duncan's murder by Macbeth. (pp. 99, 100, 115, 117, 119, 126, 127, 129.)

Nairn (4661) is a bright, up-to-date, fashionable watering-place, often styled the "Brighton of the North." The climate is dry and bracing. The fine sandy beach affords excellent bathing. Golf is known to have been played in the county as far back as 1672. In 1797 the magistrates of Nairn let the grazing of the links for three years, a condition of the lease being that it was "not to prohibit the gentlemen of the town or others from playing golff or walking on the whole Links." The present course is distant about a mile from the station and is over three and a half miles in length. There are also two nine-hole courses close to the town. In all probability Nairn was made a burgh by David I. The oldest charter now possessed by the town was granted by James VI in 1597. Nairn Castle—no trace of it remains—which was strengthened, if not erected, by William the Lyon, was one of the twenty-three royal castles in Scotland handed over to Edward I.

There is an old story that James VI once claimed to have in his dominion a town (Nairn) so large that the inhabitants of the one end did not understand the language of those at the other end. The statement was literally true, for the Gaelic-speaking Celts lived as a community apart from the English-speaking Saxons. (pp. 12, 79, 99, 100, 101, 103, 105, 109, 110, 111, 113, 117, 120, 122, 123, 124, 128, 129, 131, 132.)

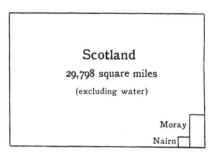

Fig. 1. Areas (excluding water) of Moray (476 sq. miles) and Nairn (163 sq. miles) compared with that of Scotland

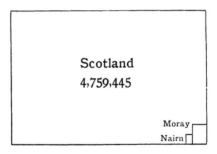

Fig. 2. Population of Moray (43,427) and Nairn (9319) compared with that of Scotland in 1911

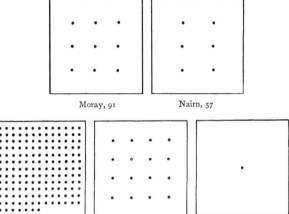

Moray, 91   Nairn, 57

Lanarkshire, 1633   Scotland, 157   Sutherland, 10

Fig. 3. Comparative density of population to the square
mile in 1911.

(*Each dot represents* 10 *persons*)

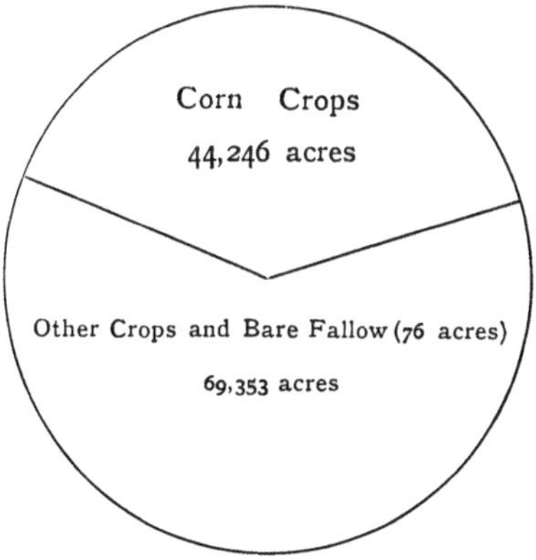

Corn   Crops

44,246 acres

Other Crops and Bare Fallow (76 acres)

69,353 acres

Fig. 4. Proportionate areas under Corn Crops compared with
that of other cultivated land in Moray and Nairn in 1912

DIAGRAMS

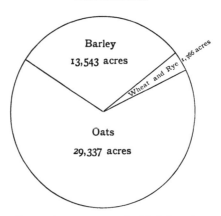

Fig. 5.   Proportionate areas of chief Cereals of Moray
and Nairn in 1912

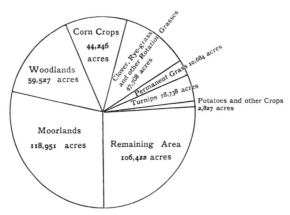

Fig. 6.   Proportionate areas of land in Moray and
Nairn in 1912

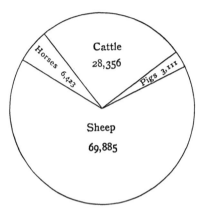

Fig. 7. Proportionate numbers of Live Stock in Moray and Nairn in 1912

www.ingramcontent.com/pod-product-compliance
Ingram Content Group UK Ltd.
Pitfield, Milton Keynes, MK11 3LW, UK
UKHW042145280225
455719UK00001B/127